Structural System Analysis

Vladimir Petrov

Structural System Analysis

Su-Field Analysis. TRIZ

 Springer

Vladimir Petrov
Tel Aviv-Yafo, Israel

ISBN 978-3-031-55827-6   ISBN 978-3-031-55825-2 (eBook)
https://doi.org/10.1007/978-3-031-55825-2

Translation from the Russian language edition: "Структурный анализ систем: Вепольный анализ. ТРИЗ " by Vladimir Petrov, © author 2018. Published by Издательские решения. All Rights Reserved.

This Springer imprint is published by the registered company Springer Nature Switzerland AG
The registered company address is: Gewerbestrasse 11, 6330 Cham, Switzerland

Paper in this product is recyclable.

Acknowledgments

I am very grateful to my teacher, colleague, and friend, Geinrich Altshuller, first of all for creating the basis of the theory of the development of technical systems—the laws of their development, and for having the good fortune to talk and discuss with him some materials from this book.

I'd like to express my deepest gratitude to my colleague, German Voronov (Israel), for co-authorship on Chaps. 7 and 8 of this book.

Introduction

Modeling is used for the analysis and synthesis systems. It is one of the components of inventive thinking or talented thinking.[1]

There are various methods of modeling, for example, real, mathematical, computer, and mental.

In this book, only mental modeling will be considered, which helps to solve complex (inventive) problems.

An inventive problem is a problem containing a contradiction, which is one of the important concepts of the theory of inventive problem solving (TRIZ).

Functional and structural analysis (Su-field analysis) are involved in modeling the structure of the system in TRIZ.

Su-field analysis is intended to represent the source system in the form of a specific (structural) model and transform it to obtain a structural solution that eliminates shortcomings.

---

[1] Vladimir Petrov. Talented Thinking. TRIZ. Springer, 2023, 221 pages, ISBN 978-3-031-15504-8, ISBN 978-3-031-15505-5 (eBook).

Contents

Chapter 1
Concepts of Su-Field Analysis

> ... **Su-field is the minimal model of a technical system**: it involves a product, tool, and energy (field) necessary for tool influence on product. Any complex technical systems can be reduced to Su-fields.
> **Genrich Altshuller**

Structured Substance-Field (Su-Field) Analysis—this section of TRIZ is intended to represent the original technical system in the form of a specific (structured) model and transform it to acquire a structured solution that avoids defects. Su-field analysis was developed by G. Altshuller.

Su-field analysis—special language to model that enables the original technical system to be presented in the form of a specific (structured) model. Features of the technical system are revealed by means of special rules. Then, the original model of the problem is converted to obtain a structured solution based on special rules that avoid defects of the original system.

Statistical analysis of technical solutions has shown that the structure of technical systems must be known in order to improve their efficiency. The model of such a structure is called the **Su-field**.

Su-field—the minimal model of a working technical system consisting of two interacting objects and their interaction.

The interacting objects are conditionally named **Substance** and are designated S_1 and S_2, and the interaction itself is called a **Field** and is denoted by **F**.

The "**substance**" encompasses any object including its material, structure, molecules, atoms; to the most complex technical systems, such as a space station. This can be a component or data in information systems.

The **field** can be any action or interaction, for example, energy, power, or information. This can be an algorithm in information systems.

Types of substances and fields are listed in the bibliography on Su-field analysis.

Su-field is expressed in model (1.1).

© The Author(s), under exclusive license to Springer Nature Switzerland AG 2024
V. Petrov, *Structural System Analysis*, https://doi.org/10.1007/978-3-031-55825-2_1

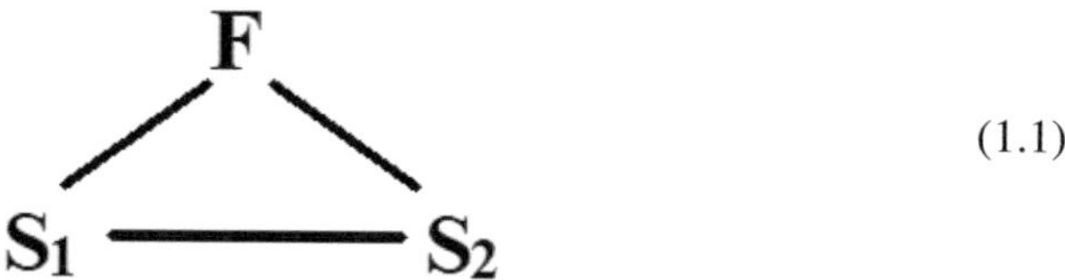

(1.1)

The term Su-field is derived from the words "substance" and "field."

The Su-field analysis includes specific rules and trends. These trends align with the law of increasing degree of Su-field which is described below.

The Su-field analysis is designed for:

- Displaying the original structure of the problem (system)
- Determination of a structured solution to the problem
- Identifying prospects of the evolution of system structure

Su-field has the form of (1.2), with S_1—product, S_2—tool "processes" product S_1, and F_1—field (energy imparted to tool).

(1.2)

Example 1. Cutting Bread

Example of Su-field model on bread cutting.

Bread (S_1) is cut with a knife (S_2), with hands applying force (F_1, mechanical field). In this case, F_1—linear movement of the knife.

This example can be presented by another Su-field model (1.3), i.e., knife (S_2) acts on bread (S_1) through a mechanical field (F_2), which shows the pressure of the knife on bread or friction between the knife and bread.

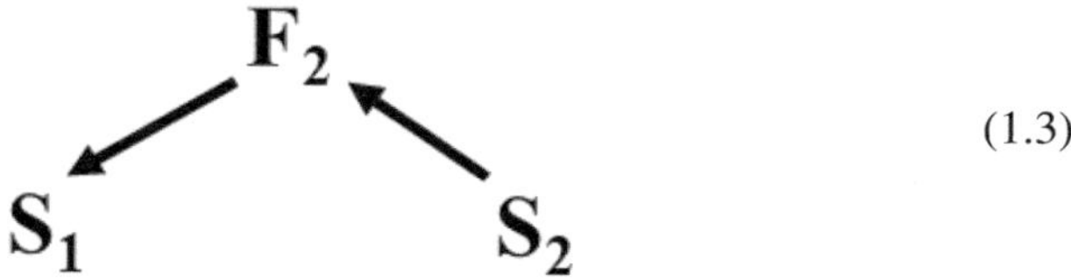

(1.3)

Example 2. Information System

Su-field model can be represented by model (1.4), if S_1—substance 1 (program), S_2—substance 2 (program), and F_1—field (signal—information). This formula can be represented as follows: S_1—data 1 (information), S_2—data 2 (information), and F_1—algorithm.

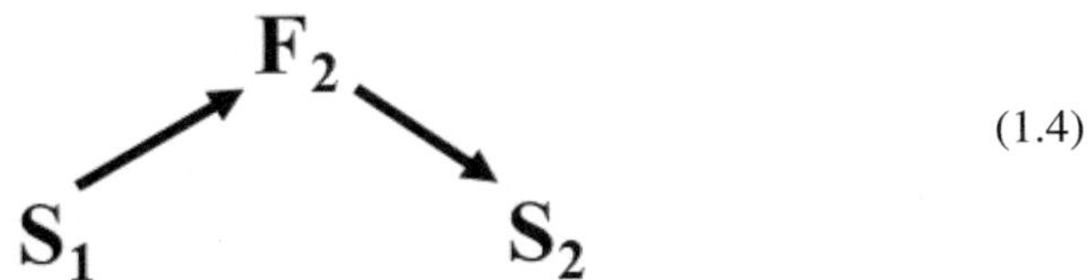

$$(1.4)$$

We introduce the concept of "responsiveness."

Responsiveness in Su-field analysis—a feature of a substance in response to the effect of field F that performs the required (given) action, or substance to generate the necessary field F.

Here are examples of "responsive" substances and fields:

1. Ferromagnetic material responsive to magnetic field
2. Resistance strain gauge responsive to deformation, pressure, voltage, displacement (mechanical field)
3. Shape memory material responsive to thermal field
4. Fluorescent and photosensitive material responsive to X-ray radiation
5. Polarizing plate responsive to optical field
6. Photodiode responsive to optical field
7. Liquid crystal responsive to thermal and electric fields

Chapter 2
Basic Configuration

This section provides the basic configuration of Su-field analysis.

The **connection** between the components is a designated line.

Model (2.1) shows the substances S_1, S_2 connected to each other in some way (not always known), and model (2.2) shows the interaction of F_1 and S_1.

$$S_1 \text{———} S_2 \tag{2.1}$$

$$F_1 \searrow \atop S_1 \tag{2.2}$$

Action is indicated by an arrow.

Impact of tool S_1 on product S_2 can be expressed by model (2.3). The arrow indicates the direction of action from S_1 to S_2.

$$S_1 \longrightarrow S_2 \tag{2.3}$$

Model (2.4) shows the action of field F_1 on substance S_1.

$$(2.4)$$

It can be the **opposite action** of S_2 on S_1 as shown in model (2.5),

$$S_1 \longleftarrow S_2 \qquad (2.5)$$

or S_1 on F_1—model (2.6).

$$(2.6)$$

Interaction designated by a two-way arrow.
Model (2.7) describes the interactions of substances S_1 and S_2.

$$S_1 \longleftrightarrow S_2 \qquad (2.7)$$

Model (2.8) describes the interaction of field F_1 and substance S_1.

$$\Pi_1 \searrow B_1 \qquad (2.8)$$

Actions can be **ineffective** or **insufficient**. These are indicated by a dashed line as shown in models (2.9) and (2.10).

$$S_1 - - - S_2 ; \quad S_1 - - \rightarrow S_2 ; \quad S_1 \leftarrow - \rightarrow S_2 \qquad (2.9)$$

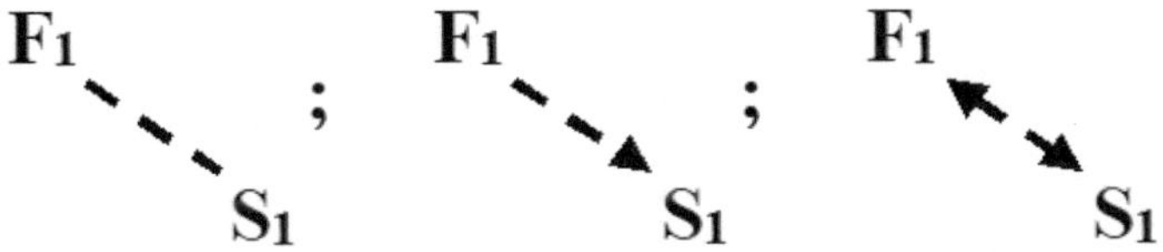

$$F_1 \quad ; \quad F_1 \quad ; \quad F_1 \qquad (2.10)$$

Excessive actions are indicated by two parallel lines (arrows). These actions are shown in models (2.11) and (2.12).

$$S_1 \Longleftrightarrow S_2 \;;\; S_1 \Longrightarrow S_2 ;\; S_1 \Longleftrightarrow S_2 \qquad (2.11)$$

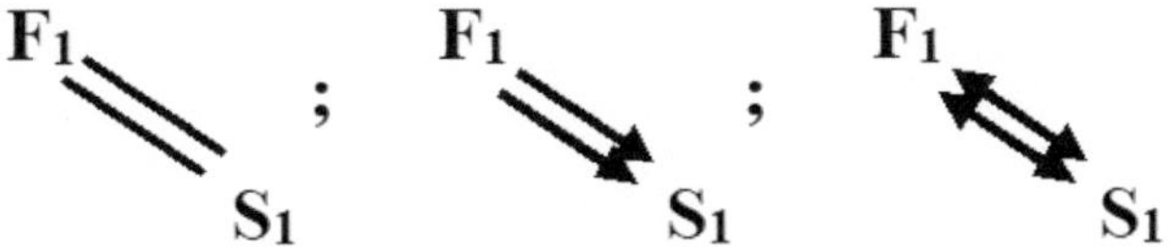

$$F_1 \quad ; \quad F_1 \quad ; \quad F_1 \qquad (2.12)$$

Harmful, **unwanted actions** are indicated by a wavy line. These are shown in models (2.13) and (2.14).

$$B_1 \sim\!\!\sim B_2 ;\; B_1 \sim\!\!\sim B_2 ;\; B_1 \sim\!\!\sim B_2 \qquad (2.13)$$

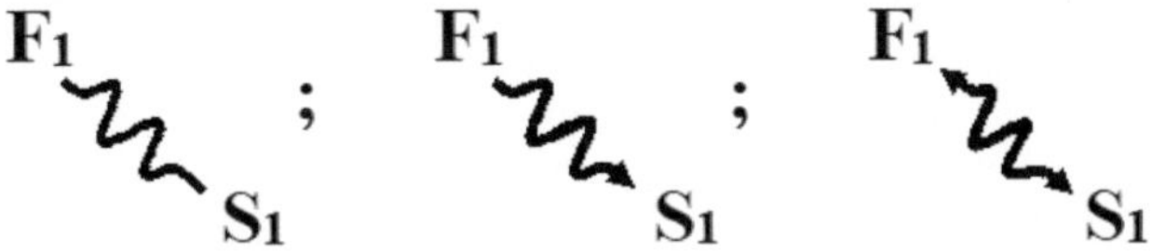

$$F_1 \quad ; \quad F_1 \quad ; \quad F_1 \qquad (2.14)$$

Transition from the initial to the final Su-field model is denoted by a double arrow, as shown in (2.15)

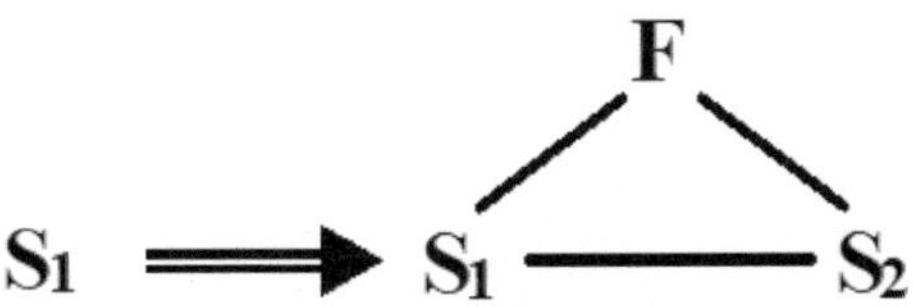

$$S_1 \Longrightarrow S_1 \stackrel{F}{\text{---}} S_2 \qquad (2.15)$$

Chapter 3
Types of Su-Field Systems

3.1 Su-Field Model for Fields

There are three types of Su-field models:

- Generation of field
- Conversion of field
- Modification of field

Generating a field from a substance is shown in model (3.1). This model describes the phenomena that occur, for example, in magnets, radioactive substances, radio, electret (electrical analog of a permanent magnet), batteries, and substances with odor.

$$\mathbf{S} \searrow \mathbf{F} \tag{3.1}$$

Substances and fields can be letters or alphanumeric instead of numerals, for example, a magnet in model (3.1) can be designated as S_{mag}, F_{mag} or S_1, F_{mag} (S_{mag}, F_1); radioactive material–$S_{ra.}$, $F_{ra.}$; radio–S_{rad}, F_{rad}; electret–$S_{el,}$ F_{el}.

Here is an example in the field of information systems.

Example 3. Correcting Codes

Data generation or transmission errors occur due to interference during data entry. Correcting codes are used for detection and error correction.

Additional information (check digit) is included during writing or transmitting of useful data. This check digit is used to detect and correct errors during reading or receiving of data.

It can be used, for example, in the detection of computer viruses by checking the determined checksum.

© The Author(s), under exclusive license to Springer Nature Switzerland AG 2024

V. Petrov, *Structural System Analysis*, https://doi.org/10.1007/978-3-031-55825-2_3

It is necessary to check S_1—left part of model (3.2). Check digit is added to S_2 during writing. The checksum F_1 determines the correctness of data S_1 (no error or virus).

$$S_1 \implies S_1 \longrightarrow S_2 \searrow F_1 \tag{3.2}$$

where

S_1 data 1
S_2 data 2 (additional information - check digit)
F_1 checksum

Substance converts a field as represented in model (3.3). The substance converts one type of field (energy or information) F_1 to another type F_2. These are two qualitatively different fields.

$$\begin{array}{c} F_1 \\ \nwarrow \\ S \\ \searrow \\ F_2 \end{array} \tag{3.3}$$

Note: It is acceptable to have the input field (in this case F_1) placed above substance S, and the output field F_2 placed below substance S.

Energy conversion occurs in, for example, a generator, engine, transformer, amplifier, or measuring component (sensor).

Example 4. Generator

A generator of electric current converts rotational field F_1 (field of mechanical power) which can be depicted as F_{mech}, into electrical field F_2 or F_{el}. Su-field has the form of (3.4).

$$\begin{array}{c} F_{mech} \\ \nwarrow \\ S \\ \searrow \\ F_{el} \end{array} \tag{3.4}$$

Example 5. Electric motor

Electric motor—reverse conversion—electric field F_{el} is converted into mechanical field F_{mech}. Su-field has the form (3.5).

$$(3.5)$$

Converting Information

Example 6. Telephone

Telephone—sound information (acoustic field F_{ac}) is converted into electrical F_{el}, and the inverse transformation—acoustic field F_{ac} into electrical F_{el}. These conversions are performed in microphones and earphones, respectively. Radio converts electromagnetic waves (electromagnetic field $F_{el.m}$) into sound (acoustic field F_{ac}).

Modification of field by substance is shown in model (3.6). The substance changes the parameters of the same field (energy or information) from F_1 to F_2. The field does not qualitatively change, so the field can be represented as F', F'', then the Su-field model (3.3) can be presented in the form (3.6).

$$(3.6)$$

The modification of energy can be achieved using, for example, transformer, transistor, amplifier, rectifier, frequency converter, analog-to-digital converter (analog-code converter), prism, and lens.

Modification of information can be achieved using, for example, code converters, information converters (e.g., decimal to binary and vice versa), and computers.

3.2 Types of Su-Field Systems for Measurement and Detection

There is a class of problems in which it is necessary to measure some parameters of the system or to detect some objects or their parts. Conventionally, such technical systems are called **measurement**. Models of such systems can have Su-field models discussed earlier (3.1), (3.3), or (3.6).

In order to measure or detect a parameter of substance S_1, it is combined with substance S_2, which can:

– generate field F_1 (3.7)
– convert field F_1 to field F_2 (3.9)
– modify field F' to field F'' (3.10)

Field generation

It is necessary to measure or detect an object, which is denoted as substance S_1. It is combined with substance S_2 which generates a field F_1.

A generated field in the Su-field model is described in model (3.7). The double arrow on the left shows that the system is necessary to detect or measure (substance S_1), and the one on the right—Su-field model of the generated field, where S_2—substance generator, which we have considered earlier.

It is possible to readily detect S_1 or measure its features using a selected field.

$$S_1 \implies S_1 \longrightarrow S_2 \searrow F_1 \tag{3.7}$$

Example 7. Finding a sunken object

In order to mark the location of a sunken object S_1, a radio buoy for rescue operations S_2 is attached, giving a signal F_{rad} (Fig. 3.1).

$$S_1 \implies S_1 \longrightarrow S_2 \searrow F_{rad} \tag{3.8}$$

where:

S_1 sunken object
S_2 radio buoy
F_{rad} radio (radio field—electromagnetic field)

Conversion of Field

It is necessary to detect substance S_1. In order to achieve this, substance S_2 which is exposed to field F_1, combines with F_1 and converts it to F_2. Conversion of field is described in Su-field (3.9).

$$S_1 \implies S_1 \longrightarrow S_2 \begin{array}{c} \nearrow F_1 \\ \searrow F_2 \end{array} \tag{3.9}$$

Note

Note that if the object of measurement, S_1, has an available field F_1 that can adequately respond (generate response field F_2), there is no necessity to add another substance S_2.

Fig. 3.1 Detection of
sunken object

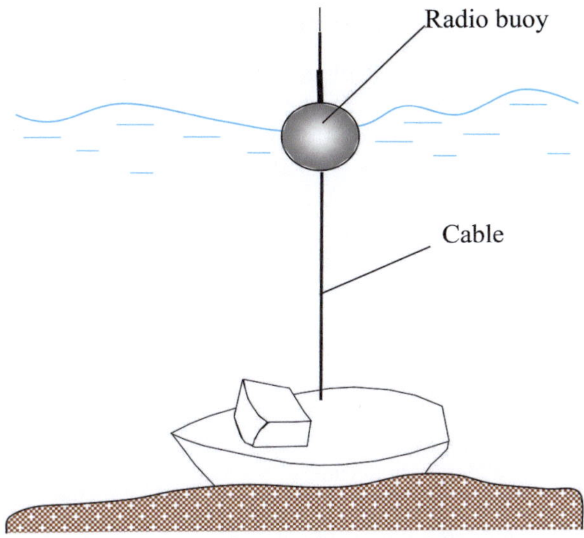

Example 8. Temperature measurement

Thermometer can be represented as Su-field (3.9).

S_1 object where the temperature is to be measured
S_2 thermometer "converting" temperature (thermal field F_1 or F_{temp}) into a certain
signal (field F_2, e.g., electrical F_{el}, or optical signal F_{opt} – mercury column),
which can be viewed.

Thus, it is possible to introduce any sensor, for example, to measure: pressure, velocity, displacement, position, strain, flow rate, humidity, and radioactivity.

Modification of field

It is necessary to detect substance S_1. In order to achieve this, it is combined with substance S_2 which influences field F'. Substance S_2 modifies it into field F". Modification of the field is described in Su-field (3.10). Fields F' and F" are of the same nature, for example, they may differ quantitatively.

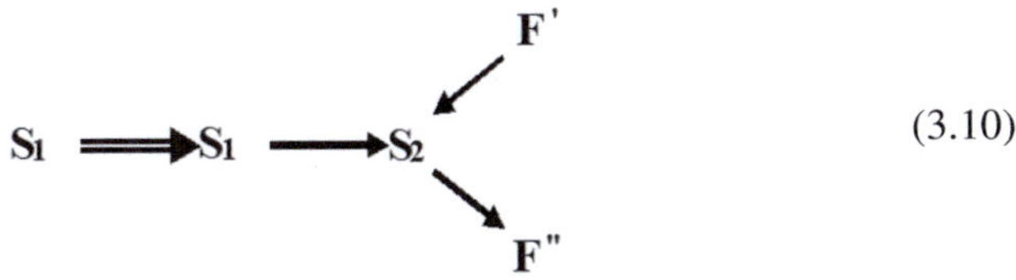

$$(3.10)$$

Su-field (3.10) can represent, for example, any electrical measurements: voltage, current, power, and frequency.

Example 9. Pedestrian detection

It is necessary to detect and not knock down a pedestrian (S_1) during the night. So, his clothes, footwear, or bag need to have reflective material (S_2). Headlights (F') of a vehicle are reflected off from substance (S_2), and the driver sees the reflected light (F"). This can be represented in Su-field (3.11).

$$S_1 \Longrightarrow S_1 \longrightarrow S_2 \diagup^{F'_{oopt}}_{\diagdown F''_{oopt}} \tag{3.11}$$

where:

S_1 pedestrian
S_2 reflective material
F'_{opt} headlights (optical field)
F''_{opt} reflected light (optical field).

Example 10. Living sensors at your fingertips

Engineers and biologists at MIT have teamed up to design a new "living material"— a tough, stretchy, biocompatible sheet of hydrogel injected with live cells that are genetically programmed to light up in the presence of certain chemicals.

The researchers demonstrate the potential of the new material for sensing chemicals, both in the environment and in the human body.[1]

The team fabricated various wearable sensors from the cell-infused hydrogel, including a rubber glove with fingertips that glow after touching a chemically contaminated surface, and bandages that light up when pressed against chemicals on a person's skin (Fig. 3.2).[2]

3.3   Types of Su-Field Models

The following are the types of structures:

- **Incomplete su-field System** (3.12), (3.13), (3.14)
- **Su-Field system–Simple su-field** (3.15)
- **Complex su-field:**

    – Internal Complex Su-field (3.19), (3.20)

---

[1] Xinyue Liu and others. Stretchable living materials and devices with hydrogel–elastomer hybrids hosting programmed cells. PNAS. February 15, 2017, URL: http://www.pnas.org/content/early/2017/02/14/1618307114.full?sid=43417b52-d915-4e2a-ab1b-610afd15fe12.

[2] Living sensors at your fingertips. URL: http://news.mit.edu/2017/living-sensors-your-fingertips-0215.

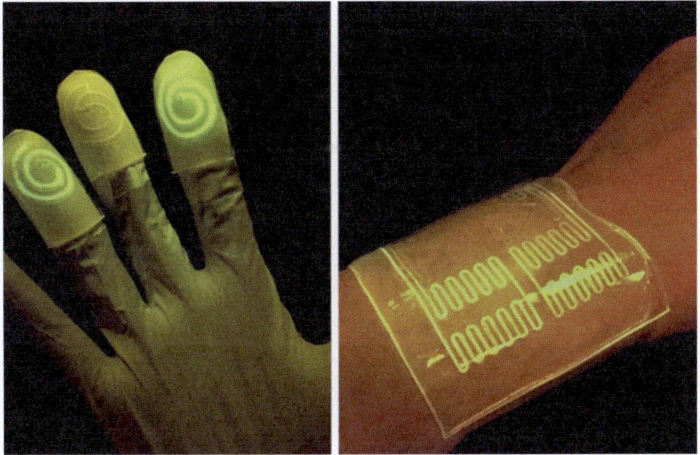

Fig. 3.2 Living sensors at your fingertips

  – External Complex Su-field (3.24), (3.25)
  – Complex Su-field on the external environment (3.29), (3.30)
  – Complex Su-field on the modified external environment (3.33), (3.34)

- **Compound Su-Field:**

  – Chain Su-field (3.39)
  – Double Su-field (3.43)
  – Mixed Su-field (3.46), (3.47)

Incomplete Su-Field system

The system consisting of one component: substance S_1 or field F_1, described by model (3.12) or two components: two substances, S_1, S_2 (3.13); substance S_1 and field F_1 (3.14), are considered as Incomplete Su-field or Incomplete Su-field system.

$$S_1;\ F_1 \tag{3.12}$$

$$S_1 \longrightarrow S_2 \tag{3.13}$$

$$\begin{array}{c} F_1 \\ \searrow \\ S_1 \end{array} \tag{3.14}$$

As a rule, an Incomplete Su-field system is uncontrolled or poorly managed.

Basic rule of Su-Field analysis

An Incomplete Su-field system needs to increase controllability in order to achieve a Su-field system. This rule can be roughly represented in the form (3.15).

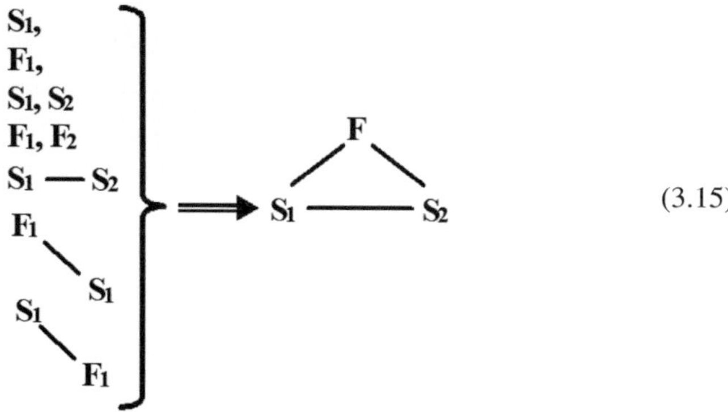

$$\left. \begin{array}{l} S_1, \\ F_1, \\ S_1, S_2 \\ F_1, F_2 \\ S_1 - S_2 \\ F_1 \searrow S_1 \\ S_1 \searrow F_1 \end{array} \right\} \implies S_1 - \overset{\displaystyle F}{\triangle} - S_2 \tag{3.15}$$

Problem 1. Removing bark from wood

Conditions of problem

Normally, wood bark is separated mechanically in special debarking drums or mechanical tools, for example, with an axe. However, the wood is damaged.

It is necessary to provide a method of separating bark from wood, without spoiling the wood.

Analysis of problem

Consider this problem from the standpoint of Su-field analysis. There are wood and bark. An Incomplete Su-field system. It can be described by model (3.16).

$$S_1 \longleftrightarrow S_2 \tag{3.16}$$

where:

S_1 wood
S_2 bark

It is necessary for the system to be converted into a Su-field. Complete Su-field is achieved by introducing field F_1, which acts only on the bark in the direction of separation from the wood. This Su-field is shown by model (3.17).

$$S_1 \longleftrightarrow S_2 \implies S_1 \overset{\displaystyle F_1}{\nearrow} \longrightarrow S_2 \tag{3.17}$$

It is necessary to choose field F_1 which can carry out such an action.

There is a layer of cells (cambium) between bark and wood (Fig. 3.3) that contains a large amount of moisture. This layer allows the wood to tear off from the bark

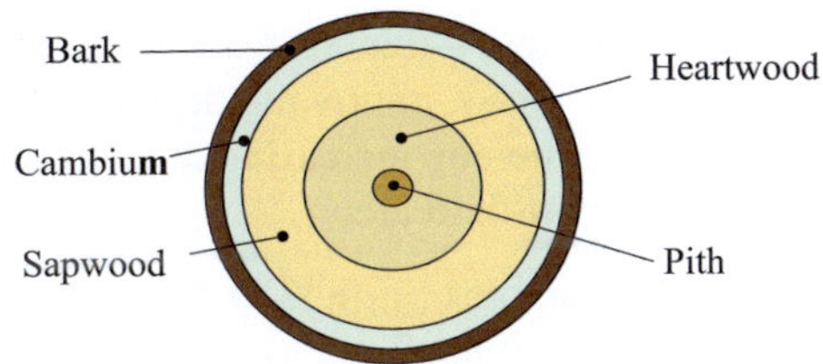

Fig. 3.3 Cross-sectional view of tree trunk

through boiling. Boiling can be accomplished by vacuuming or heating, e.g., high-frequency currents. Thus, Su-field analysis recommends the use of a thermal field $F_1 = F_{therm}$.

Problem 2. Tracking an object

Conditions of problem

It is necessary to track an object, S_1.

Analysis of problem

Given only one substance S_1—object to be tracked. The system is an Incomplete. Su-field.

Complete the Su-field system in order to obtain a solution. This type of problem is on measurement and detection. Use model (3.2).

In order to track the object, a "bug" is attached. The location of the tracked object is determined with the help of special equipment.

Thus, we have an object to be tracked, S_1. We construct a Su-field model of the object being tracked. In order to achieve this, we add one more substance S_2 ("radio bug") that generates a field F_1 (radio field).

Su-field has the form of (3.18).

$$S_1 \implies S_1 \longrightarrow S_2 \searrow F_1 \tag{3.18}$$

where:

S_1 object being tracked
S_2 "radio bug"
F_1 radio field

A further increase in controllability of the Su-field system is by replacing substances and/or fields, and changing the Su-field model.

Consider the types of Su-field models.

As noted above, Su-field models can be complex and compound. Let's consider these structures.

Complex Su-Field

Complex Su-field—this Su-field has an additional incorporated substance S_3 that can bind to S_1 or S_2, improving the controllability of the system or providing it with new features, thereby increasing the efficiency of the technical system.
  Complex Su-field are:

– Internal Complex Su-field (3.19) and (3.20)
– External Complex Su-field (3.24) and (3.25)
– Complex Su-field on the external environment (3.29) and (3.30)
– Complex Su-field on changed external environment (3.33) and (3.34)

Internal Complex Su-Field – this complex Su-Field is where additive S_3 is attached to substance S_1 (3.19) or S_2 (3.20). The introduction of internal substance is shown in the form of brackets.

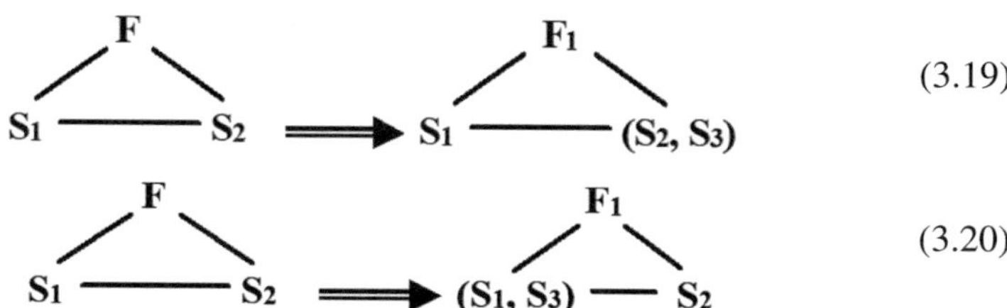

$$(3.19)$$

$$(3.20)$$

Problem 3. Clean up oil spills

Conditions of problem

Oil spills on the sea surface are a result of tanker accidents.
  One of the methods for cleaning up oil spills on the water surface is as follows. The oil spill is surrounded by floating barriers that prevent spreading. Then, it is filled with granules that are porous–absorbents that absorb oil.
  The problem is in the collection of the granules that have been soaked with oil.

Analysis of problem

There are granules S_1, soaked with oil S_2.
  The system is an Incomplete Su-field. It is represented by model (3.21). Granule S_1 acts on oil S_2 through absorption (capillary effect).

$$S_1 \longrightarrow S_2 \qquad (3.21)$$

where.

S_1 granule
S_2 oil

We need to complete the Su-field system in order to solve the problem. It is necessary to find a field that is responsive to granules with oil, that can be easily removed. Such a field is difficult to find, so add another substance S_3 to granules S_1

that will respond to the incorporated field F_1. This field must remove the granule, along with itself and the oil.

It is proposed to add ferromagnetic particles S_3 to the granules so that they will be easy to collect using magnetic field F_1.

Su-field model (3.22).

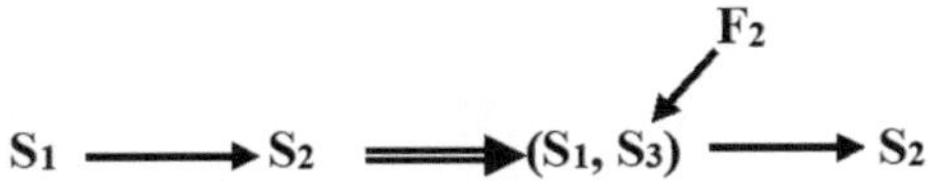

$$S_1 \longrightarrow S_2 \Longrightarrow (S_1, S_3) \overset{F_2}{\longrightarrow} S_2 \tag{3.22}$$

External Complex Su-Field—Complex Su-field where additive S_3 is attached externally to S_1 (3.24) or S_2 (3.23). This type of complex Su-field is used when it is impossible or undesirable to introduce S_3 into the existing substance.

$$\overset{F}{\triangle}\; S_1 \text{—} S_2 \Longrightarrow \overset{F_1}{\triangle}\; S_1 \text{—} S_2, S_3 \tag{3.23}$$

$$\overset{F}{\triangle}\; S_1 \text{—} S_2 \Longrightarrow \overset{F_1}{\triangle}\; S_1, S_3 \text{—} S_2 \tag{3.24}$$

Problem 4. Removal of radio components

Conditions of problem

Removal of radio components is by means of a soldering iron. However, radio components are often overheated (thermal shock) which leads to damage.

What can be done?

Analysis of problem

Construct a Su-field model of the described system. It can be shown by model (3.25).

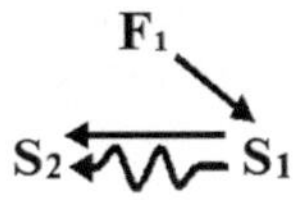

$$\overset{F_1}{\searrow}\; S_2 \rightleftharpoons\!\!\wedge\!\!\wedge\!\!- S_1 \tag{3.25}$$

where.

F_1 thermal field from the heated soldering iron
S_1 tin
S_2 pin (leg) of radio component

The problem is described using Su-field with useful and harmful interactions. Useful action (straight line arrow from S_1 to S_2)—tin melts and releases the pin of the radio component. Harmful action (wavy arrow from S_1 to S_2)—hot tin overheats the pin of the radio component.

Model (3.25) does not reflect the full picture of the problem. A more detailed type of Su-field model can be represented by (3.26).

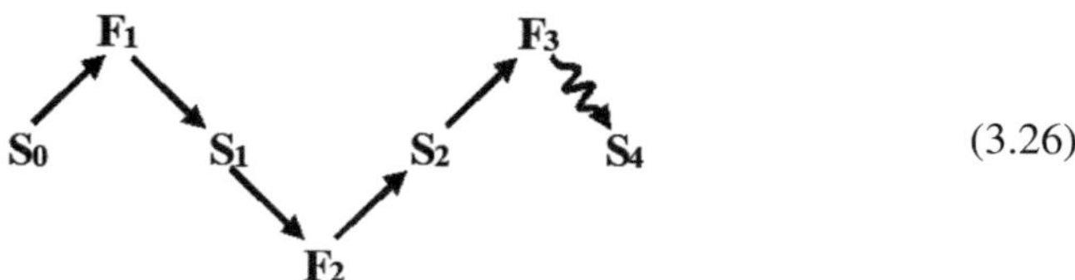

$$(3.26)$$

where.

S_0 tip of soldering iron
F_1 thermal field of heated soldering iron
S_1 tin
F_2 thermal field of heated tin
S_2 pin (leg) of radio component
F_3 thermal field of preheated pins of radio component
S_4 radio component

This model is certainly more accurate, but more complex. It does not have any heuristic power but allows focus on additional components instead of only on the main undesirable effect. In this regard, we recommend the use of a simpler Su-field model. In this case—model (3.25).

One possible solution is to go to an External complex Su-field (3.27), i.e., need to introduce an additional external substance labeled as S_3.

$$(3.27)$$

In order to prevent the radio component from being damaged by thermal shock during de-soldering, S_3 with a melting point below the melting point of the solder, is introduced into the solder (Fig. 3.4). The additional solder is an alloy of tin–lead–bismuth that significantly reduces thermal shock on radio components.

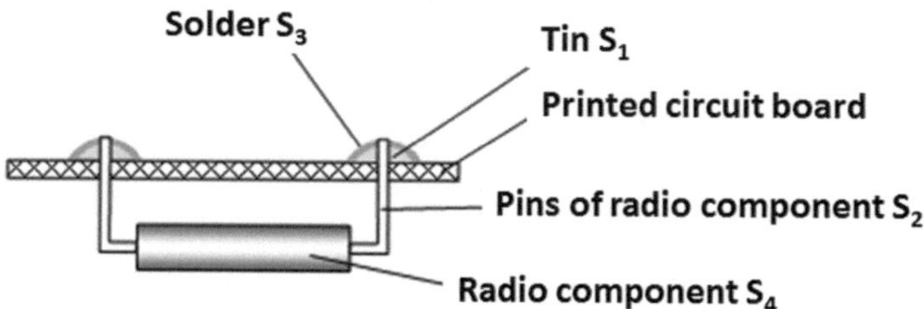

Fig. 3.4 Introduction of low-temperature solder

Complex Su-Field on external environment – External complex Su-Field where S_3 is applied on external environment S_{EE} that can be added to S_2 (of external environment, S_3 that can be added to S_2 (3.28) or S_1 (3.29).

This type of Complex Su-Field should be used when it is impossible or undesirable to attach S_3 to available substances in the system.

$$\begin{array}{ccc} & F_1 & & & F_2 \\ S_1 \longleftarrow S_2 & \Longrightarrow & S_1 \longleftarrow S_2,\ S_{EE} \end{array} \tag{3.28}$$

$$\begin{array}{ccc} F & & F_1 \\ S_1 \longrightarrow S_2 & \Longrightarrow & S_1,\ S_{EE} \longrightarrow S_2 \end{array} \tag{3.29}$$

S_{EE}—substance of external environment, S_3.

Problem 5. Clearing railway tracks

Conditions of problem

Clearing of railway tracks is carried out using a special locomotive or outboard equipment. This is not ideal as it is necessary to acquire specialized equipment, and spend extra energy, time, and human resources; on operation and maintenance.

How to avoid this?

Analysis of problem

Su-field model of the problem is in the form of (3.30).

$$\begin{array}{c} F_1 \\ S_1 \longleftarrow S_2 \end{array} \tag{3.30}$$

where.

S_1 dirt or snow
S_2 brush
F_1 brush rotation

One possible solution is to move to a Complex Su-field in an external environment (3.31).

$$\begin{array}{ccc} & F_1 & & & F_2 \\ S_1 \longleftarrow S_2 & \Longrightarrow & S_1 \longleftarrow S_3,\ S_{EE} \end{array} \tag{3.31}$$

where.

S_1 dirt or snow
S_2 brush
F_1 brush rotation
S_3 reflector
S_{EE} air
F_2 incoming flow

It is possible to clean railway tracks by running a locomotive with an airflow, which is directed at the desired place by means of special screens and apertures (Fig. 3.5). Each locomotive can be equipped with such a device (Inventor's certificate 1,054,483). It can be installed during locomotive manufacturing. Then railway tracks do not need to be specially cleaned.

In this invention, resources used–incoming airflow.

Complex Su-Field on modified external environment—this External Complex Su-field is where S_3 can be added to S_1 (3.33) or S_2 (3.32), in order to modify external environment S'_{EE}.

This change can be understood as the expansion of the external environment in forming components and additives in the external environment.

This type of Complex Su-field should be used when it is impossible or undesirable to attach S_3 to existing substances in the system or external environment.

$$
\begin{array}{ccc}
\overset{\textstyle F}{\diagup\;\diagdown} & \Longrightarrow & \overset{\textstyle F_1}{\diagup\;\diagdown} \\
S_1 \longrightarrow S_2 & & S_1 \longrightarrow S_2, S'_{BC}
\end{array}
\tag{3.32}
$$

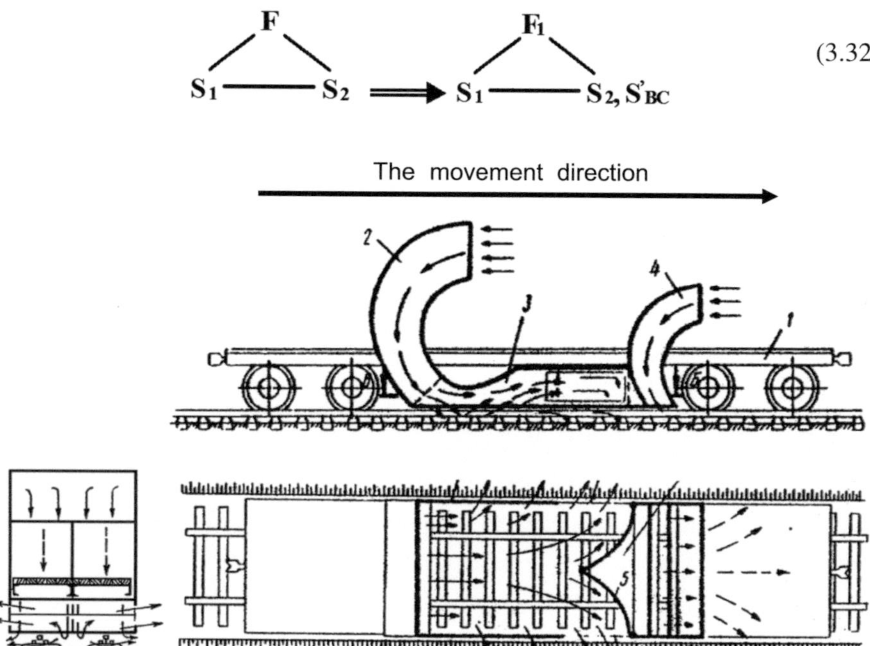

Fig. 3.5 Cleaning of railway tracks. Inventor's certificate 1,054,4831—carriage, 2–4—ducts, 2—intake air duct, 3—guide duct, 4—auxiliary duct, 5—front wall of duct, 6—side walls of duct, 7—exhaust ports.

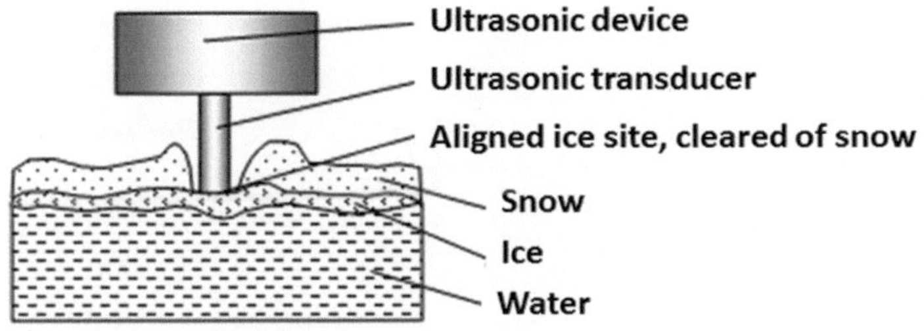

Fig. 3.6 Measuring the depth of a river

S'_{EE}—modified substance of external environment, $S_3 = S'_{EE}$.

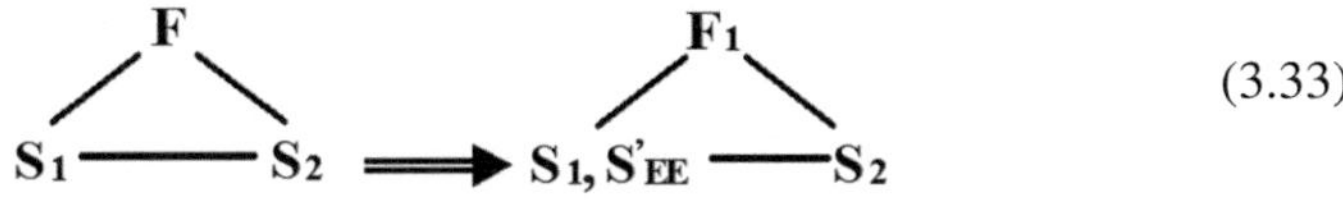

$$\tag{3.33}$$

Problem 6. Measure the depth of a river

Conditions of problem

When measuring the depth of a river through its icy surface, it is necessary to ensure reliable contact of the ultrasonic transducer with ice. There is snow on the ice surface, which needs to be cleared beforehand. Ice has an uneven surface which makes transducer contact with ice only in certain areas. The transducer needs to be aligned in order to improve contact with ice (Fig. 3.6). This is laborious and time-consuming.

What can be done?

Analysis of problem

Su-field model of the problem can be represented in the form of model (3.34).

$$\tag{3.34}$$

where.

S_1 ice
S_2 ultrasonic transducer
F_1 ultrasonic field

One possible solution is to transition to a Complex Su-field on a modified external environment (3.35).

Fig. 3.7 Compacting of
snow

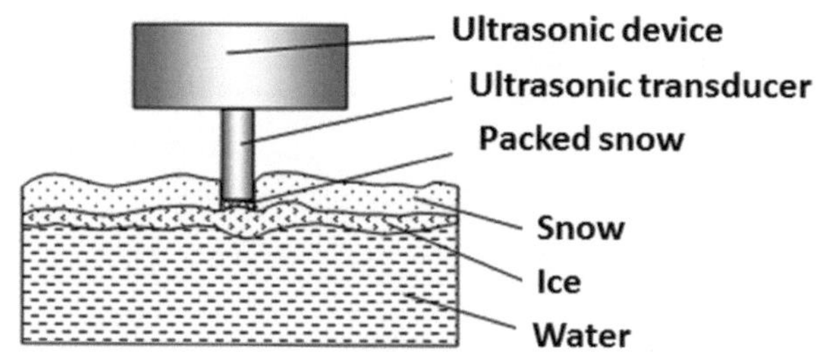

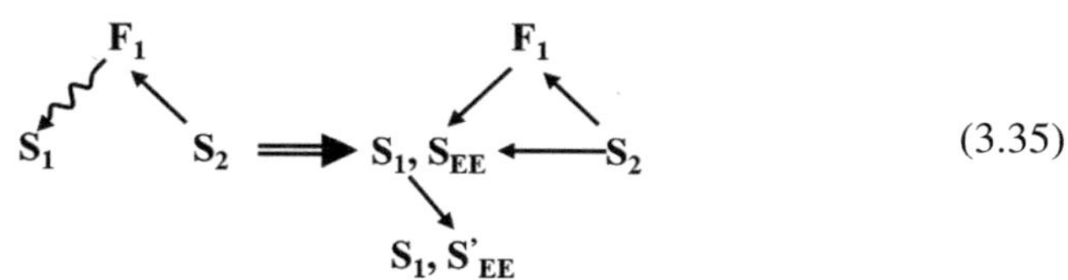

$$(3.35)$$

where.

S_1 ice
S_2 transducer
F_1 ultrasound
S_3 snow
S'_{EE} compacted snow

The intimate contact between the transducer and ice can be achieved if snow is
compacted with the help of a transducer (Inventor's certificate 900,233).
Resources used—snow (Fig. 3.7).

Complex Su-Field—this Su-field model is formed by a combination of Simple
Su-field models (3.36) and (3.37).

$$(3.36)$$

$$(3.37)$$

Compound Su-field can be divided into three types:

- Chain (3.38)
- Double (3.42)
- Mixed (3.45), (3.47)

Chain Su-Field is formed by combining Simple Su-field models (3.36). The chain Su-field model is represented by model (3.38).

Chain Su-field—Complex Su-field where substance S_2 is deployed independently of Su-field including F_2, S_3, and interactions between them.

$$\tag{3.38}$$

The brackets in model (3.38) show the new Su-field deployed from substance S_2.

Example of the chain Su-field model in Problem 4 (removal of radio components) is shown in model (3.26).

Problem 7. Determination of hidden defects

Conditions of problem

How to identify hidden defects, for example, fatigue cracks in aircraft engine turbine blades?

Analysis of problem

This is a problem of detection. It is necessary to identify defects in turbine blade S_1. It is possible to select field F_1 that responds to S_1. Then, in accordance with Su-field (3.7), it is possible not to introduce S_2 to detect defects in S_1.

Su-field model to find a solution is in the form (3.39).

$$\tag{3.39}$$

The blade is agitated by a source of mechanical vibrations (inductor coil). The inductor coil which is coupled to a power amplifier becomes a generator of electrical oscillations. The frequency of the oscillator is adjusted to its resonance frequency. A microphone which is placed near the blade, transmits the vibrations in electrical form to an oscilloscope (Fig. 3.8). The changing waveforms are studied for the presence of fatigue crack.

The main point in this solution—defect is determined "by sound". The blade provides an oscillatory motion corresponding to field F_1. The described solution corresponds to Su-field (3.40) where.

F_1 field of mechanical vibrations (possible to designate F_{vib} or F_{mech})
S_1 blade
F_2 acoustic field—air oscillation (F_{ac})

This Su-field can be represented as model (3.40).

Fig. 3.8 Determination of
hidden defects

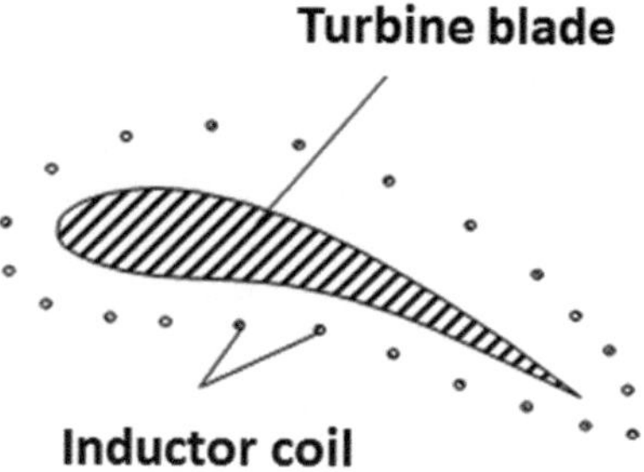

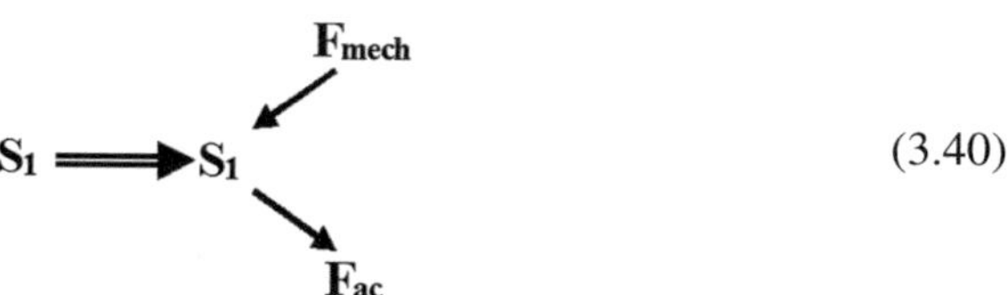

$$S_1 \Longrightarrow S_1 \quad \nearrow^{F_{mech}} \searrow_{F_{ac}} \qquad (3.40)$$

The solution can be presented in a more Complex Su-Field as described by model
(3.41).

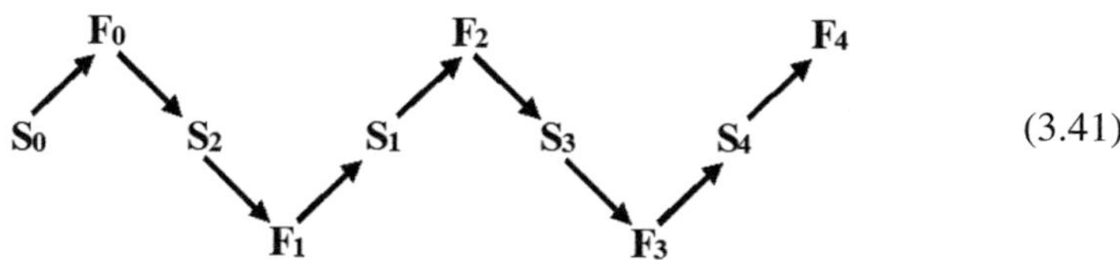

$$(3.41)$$

where:

S_0 generator of electrical oscillations
F_0 field of electrical oscillations
S_2 inductor coil
F_1 variable magnetic field (generator of mechanical vibrations)
S_1 blade
F_2 acoustic field
S_3 microphone
F_3 electrical signal
S_4 oscilloscope
F_4 light signal (image fluctuations on oscilloscope screen)

Such a Su-field is identified as a chain.
   This model can be more complicated if desired.
   Su-field (3.41) is represented by a number of different systems:

S_0, F_0 generator of electrical oscillations
S_2, F_1 electrical coil
S_3, F_3 microphone
S_4, F_4 oscilloscope

These are auxiliary systems. The main idea–the measurement of the "tone of sound" F_2, which is obtained as a result of the excitation of field F_1 by blades S_2. This solution can be implemented in other ways, for example, agitate and remove fluctuations by means of piezoelectric transducers.

As stated before, this problem of detecting (defect in blade), is carried out by measuring signal F_2, and therefore the problem is on measurement too.

Double Su-Field is formed by combining Simple Su-field models (3.36) and (3.37). The double Su-field model is shown in model (3.42).

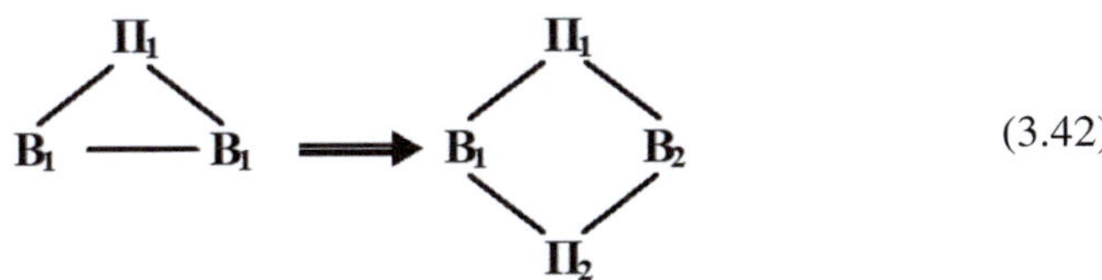

$$(3.42)$$

Problem 8. Pouring of molten metal

Conditions of problem

Pouring of molten metal S_1 from ladle S_2 is performed through an opening at the bottom (Fig. 3.9) by gravity action, F_1. Su-field model of the system is presented in the form of (3.43).

$$(3.43)$$

Such pouring is uneven as it is dependent on height h of the molten metal column (hydrostatic pressure). How to enable uniform pouring?

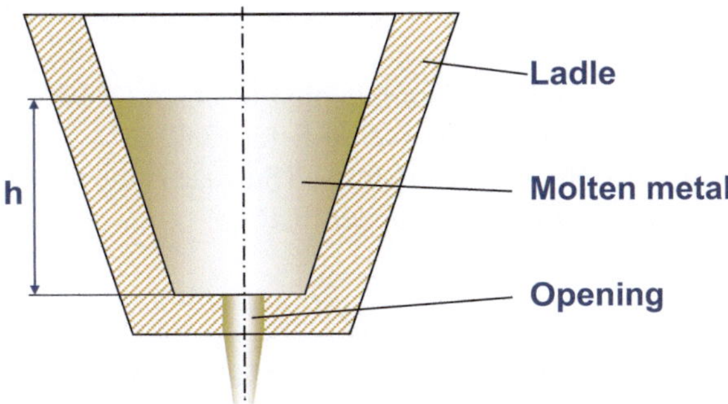

Fig. 3.9 Pouring of molten metal

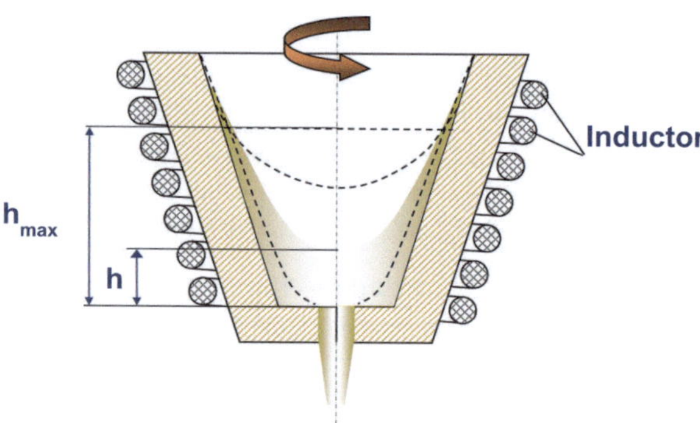

Fig. 3.10 Rotation of molten metal

Analysis of Problem

In order to enable uniform pouring, it is necessary to compensate for gravitational force, i.e. through influence from another field F_2—go to double Su-field (3.44).

Hydrostatic pressure controls the height h of the molten metal column above the opening of the casting ladle through F_2 which rotates metal in the ladle (Fig. 3.10) e.g. electromagnetic field (Inventor's certificate 275,331).

The rotation of the metal in the ladle depends on the rotational velocity of various parabolic shapes (dashed lines in Fig. 3.10). The maximum height h_{max}, is when there are no rotations (rotation velocity, $V_o = 0$). Maximum velocity of rotation (V_{max}) must correspond to a truncated parabola over the opening when metal is absent ($h_{min} = 0$) and consequently, it does not come out. Thus, it is possible to regulate metal flow through the bottom opening of the pouring ladle.

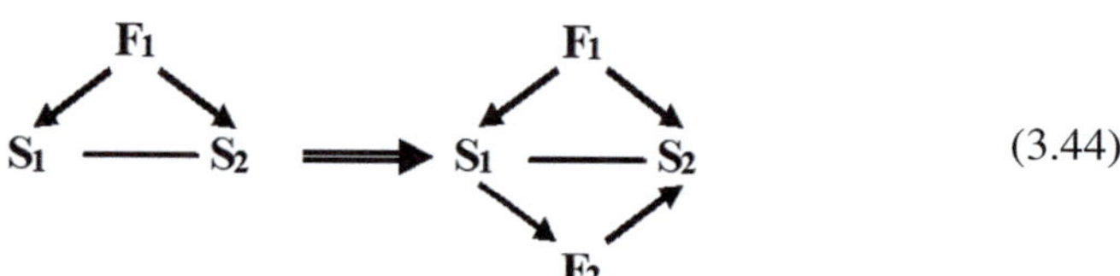

$$
\begin{array}{cc}
& F_1 \\
S_1 \text{——} S_2
\end{array}
\quad\Longrightarrow\quad
\begin{array}{c}
F_1 \\
S_1 \text{——} S_2 \\
F_2
\end{array}
\tag{3.44}
$$

Mixed Su-Field is a combination of chain (3.38) and double (3.42) Su-fields or a combination of two double Su-fields (3.42).

Transition from chain to mixed Su-field is shown in model (3.45), and the transition from double to mixed –model (3.46).

$$
\begin{array}{ccc}
F_1 & & F_2 \\
S_1 \text{——} S_2 \text{——} S_3
\end{array}
\quad\Longrightarrow\quad
\begin{array}{ccc}
F_1 & & F_3 \\
S_1 \text{——} S_2 \text{——} S_3 \\
F_2 & & F_4
\end{array}
\tag{3.45}
$$

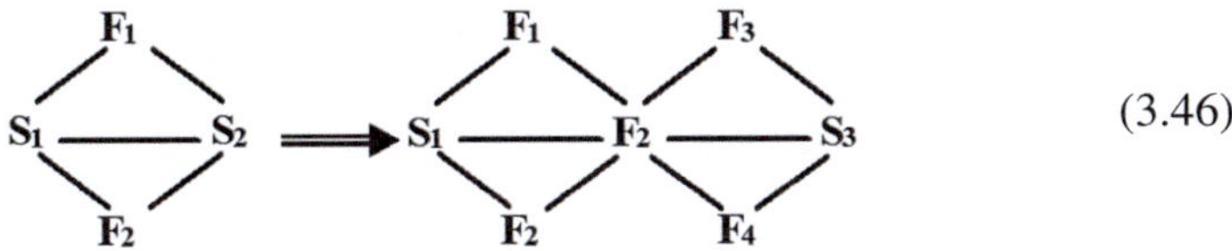

$$(3.46)$$

Example 11. Cyclones

Bulky filters are used for air purification in industrial environments. The Su-field can be represented as (3.47).

$$(3.47)$$

where:

S_1 air
S_2 dust
F_1 airflow.
S_3 filter

This is an **Internal Complex Su-field model**.

The next step in the evolution of air purification systems—use of cyclones (Fig. 3.11). Polluted air spins at high speed in a cyclone, causing particles of dust in the air to strike the walls and fall into a dust collector.

Double Su-field is used in this solution, according to model (3.48).

Fig. 3.11 Cyclone

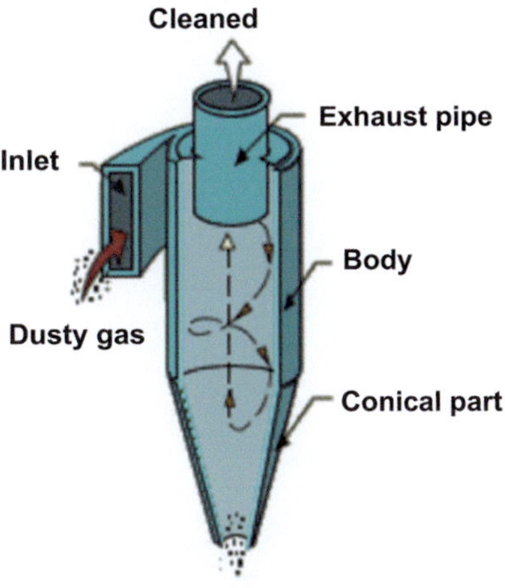

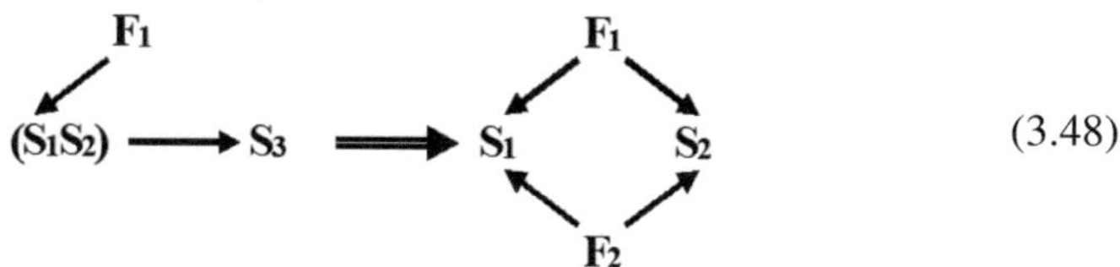

$$(3.48)$$

where:

S_1 air
S_2 dust
F_1 airflow
F_2 centrifugal force

This solution can be improved.

The disadvantage of the cyclone is that fine dust does not reach the dust collector, but settles on the walls of the exhaust pipe (extractor). Therefore, it is necessary to stop the cyclone and clean the pipe from time to time.

Let's try to go to a mixed Su-field (3.46), i.e. add F_3 to act on S_2, generating field F_4, which acts on dust S_2 (3.49).

In order to prevent the dust from clogging up the extractor, the whole pipe is transformed into an electrode – a hollow cylinder of metal, with spiked needles, located at the outlet of the pipe. The electrodes produce an electric field that repels dust from the pipe (Fig. 3.12). So, the dust ends up in the dust collector.

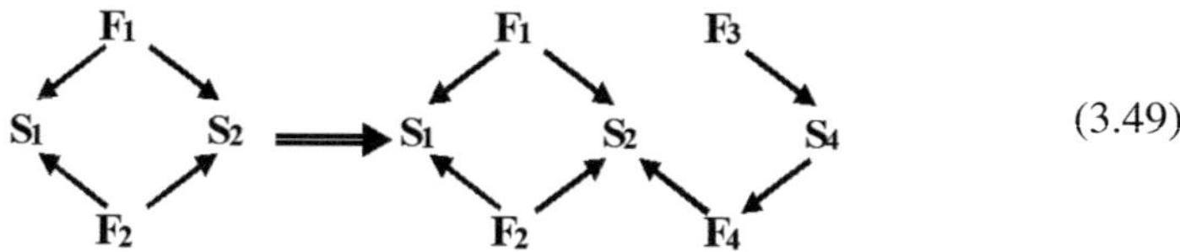

$$(3.49)$$

where:

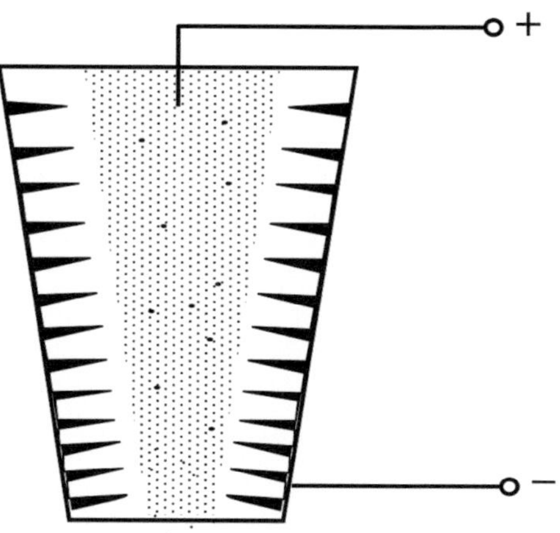

Fig. 3.12 Electrostatic field (conical part of cyclone—Fig. 3.11.)

S_1 air
S_2 dust
F_1 airflow
F_2 centrifugal forces
F_3 electric field
S_4 needles on pipe
F_4 static electricity (electric field)

Chapter 4
Elimination of Harmful Interactions

4.1 Trends of Elimination of Harmful Interactions

Quite a significant class of problems is associated with undesirable effects, which is a harmful interaction between substance and substance, field and substance; or harmful effects of field.

Remove harmful interactions by using specific rules (Figs. 4.1, 4.2 and 4.3):

1. Harmful interaction between substances (Fig. 3.12):

   - Introduction of a third substance S_3—model (4.1)
   - Introduction of a third substance S_3, which is a modification of existing substances, S_1 and S_2 ($S_3 = S_1'$, S_2') or the substance itself ($S_3 = S_1$, S_2)—model (4.4)
   - Introduction of a third substance $S_3 = S_1'$, S_2', and field F_2, that acts on S_1 or S_2, modifies S_1' or S_2'—model (4.7)

2. Harmful interaction between field and substance (Fig. 4.2)

   - "Decelerate" harmful effects—model (4.9)
   - Introduction of a second field F_2—model (4.10)
   - Introduction of a third substance S_3, which generates F_2—model (4.13)
   - Introduction of a third substance S_3, which generates F_2 under the influence of F_3—model (4.15)

3. Harmful interaction between substance and field (Fig. 4.2). Control output field by:

   - Introduction of additional substance S_2 and field F_2—models (4.18)–(4.20)
   - Change existing substance S_1 on S_2 and introduction of additional field F_3 which controls the output of field F_2—models (4.22)–(4.24).

   The purpose of the third group is to control the output field F_2.

V. Petrov, *Structural System Analysis*, https://doi.org/10.1007/978-3-031-55825-2_4

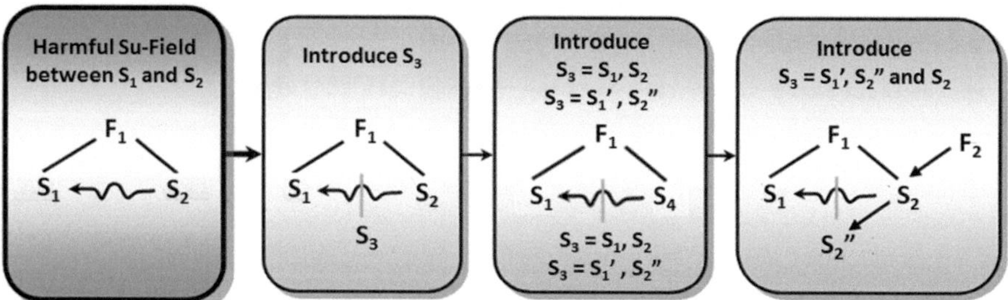

Fig. 4.1 Trend of elimination of harmful interaction between substances

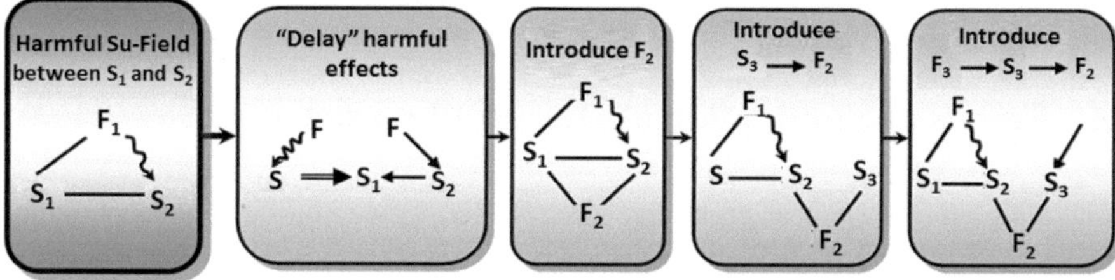

Fig. 4.2 Trend of elimination of harmful interaction between field and substance

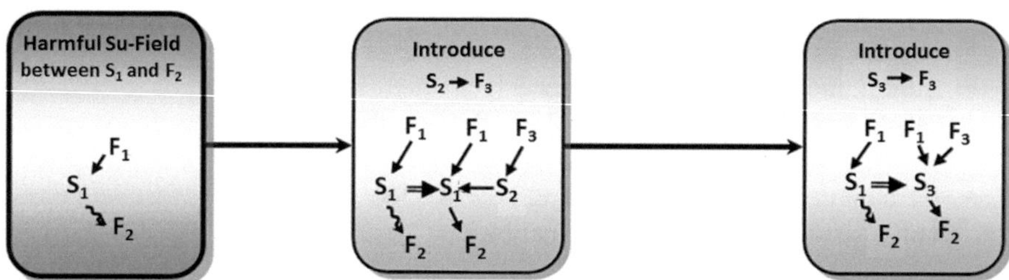

Fig. 4.3 Trend of elimination of harmful interaction between substance and field

4.2   Elimination of Harmful Interaction by Introducing S_3

Elimination of harmful interactions in the system is performed by introducing a foreign third substance S_3, between substances S_1 and S_2.

It is described by model (4.1):

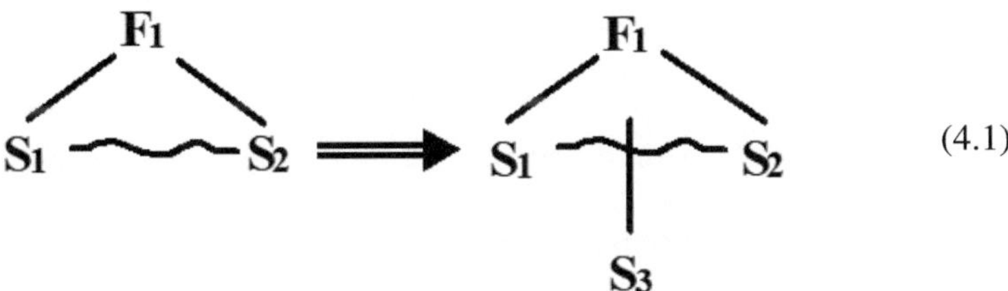

$$(4.1)$$

The introduced substance S_3 can be at the macro- and micro-level.

Problem 9. Hydrofoil

Conditions of problem

When hydrofoil is in motion, cavitation on the wing occurs which results in erosion (corrosive material). Cavities are formed and the wing loses its effectiveness (Fig. 4.4).
What can be done?

Analysis of problem

The problem is in the form of the Su-field type (4.2).

$$(4.2)$$

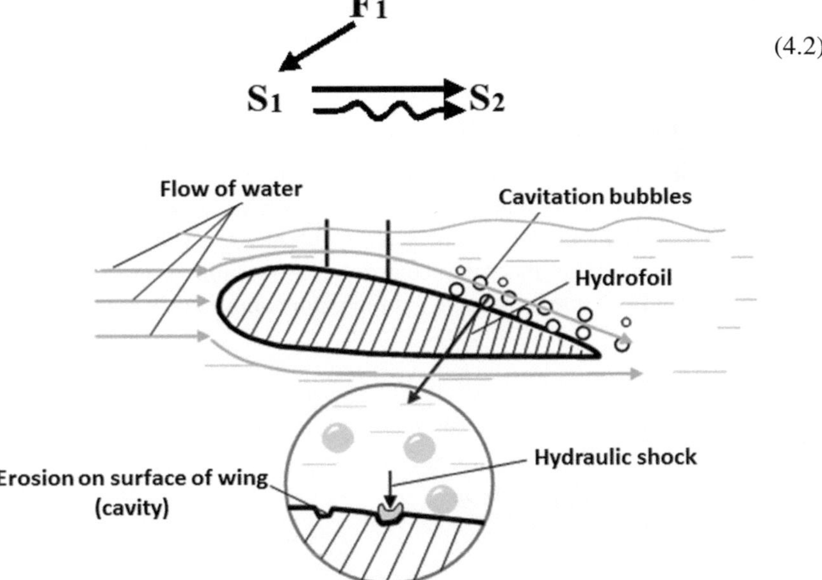

Fig. 4.4 Hydrofoil

where:

S_1 water
S_2 wing
F_1 flow of water

The flow of water acts on the wing creating lifting force (straight line arrow), and the flow of water acts on the wing forming cavities (wavy arrow—harmful action).

This Su-field has harmful interaction.

The harmful interaction can be eliminated by introducing S_3 in accordance with model (4.1). A structural solution can be represented by model (4.3) for this problem.

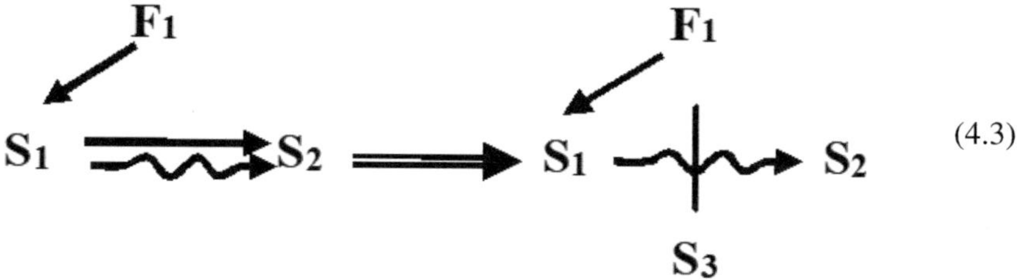

$$(4.3)$$

It is possible to use S_3:

1. Hair (Fig. 4.5)—macro S_3. They convert the turbulent flow (flow with vortices) into laminar flow (smooth—without vortices).
2. Substances with long molecules (S_3 hair at micro-level). Substances such as gels, polymers, etc. can be used. This phenomenon is called the Toms effect.

Figure 4.4: S_3—hair

Example 12. Underwater vehicle

In order to reduce the resistance of underwater vehicle, a weak solution of polymer (S_3 at micro-level) is formed at the boundary layer with seawater by heating a fluid mixture of granulated or powdered polymer with seawater. The heated fluid mixture is a dispersion of polymer macromolecules that are soluble in sea water at ambient temperature, but insoluble in water at a temperature above 70 °C. When the warmed

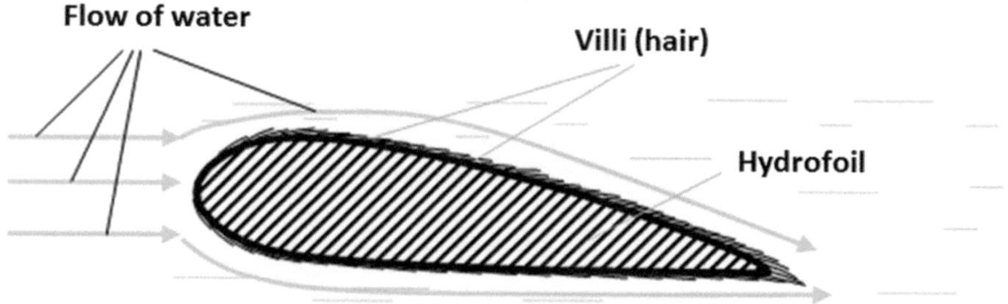

Fig. 4.5 Hydrofoil, covered with hair

fluid mixture is exposed to cold water under corresponding environmental conditions, the particles swell and dissolve, forming a sticky mass. They form a molecular solution of macromolecules in the boundary layer and prevent the formation of turbulent flow (Fig. 4.6). This invention uses the Toms effect.

Su-field model for given invention based on model (4.3):

S_1 sea water
S_2 underwater vehicle
F_1 flow of water
S_3 sticky mixture

Example 13. Pipeline

A long chain of pseudo-plastic polymer (e.g. polyacrylamide in an amount of 0.01–0.2% by weight) is introduced, in order to reduce pressure losses when moving liquid through a pipeline and to achieve features of a liquid (Fig. 4.6). This invention (Inventor's certificate 244,032) uses the Toms effect.

Su-field model for given invention based on model (4.3) (Fig. 4.7):

S_1 fluid
S_2 pipeline
F_1 fluid flow
S_3 long-chain polymer

Example 14. Reducing drag

Reduction in flow resistance can be achieved by forming additives under the influence of some fields from molecules of most fluids which have similar features to polymer molecules.

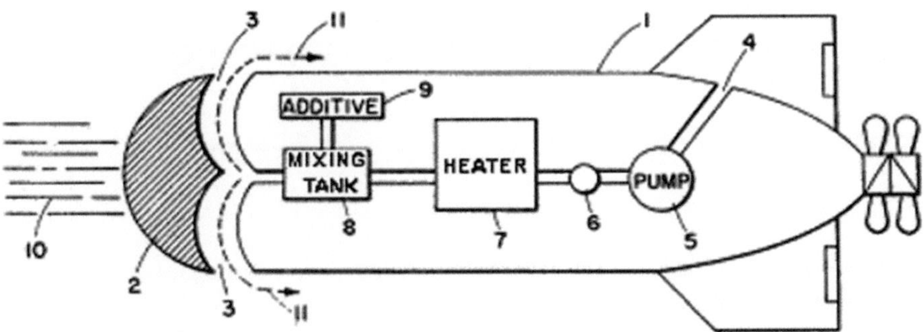

Fig. 4.6 Underwater vehicle with sticky mass. U.S. Patent 3,435,796. Where 1—underwater vehicle: 2—forming nozzle (head); 3—radial channel; 4—inlet; 5—pump; 6—valve; 7—heater; 8—mixing tank; 9—additive; 10—flow of water; 11—sticky mass—polymer dispersion (dotted line)

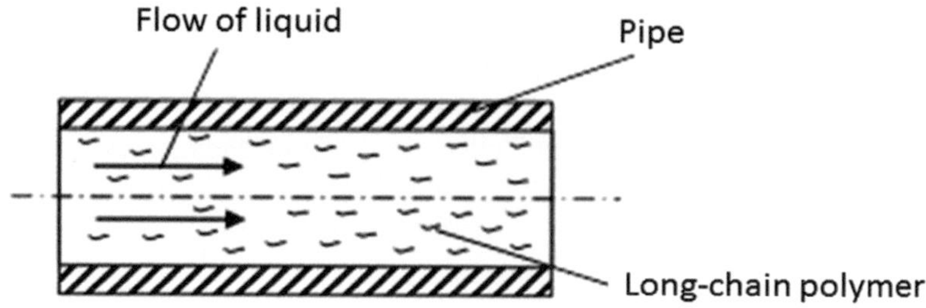

Fig. 4.7 Pipeline

In this example, S_3—additive.

4.3   Elimination of Harmful Interaction by Introducing $S_3 = S_1, S_2$ or Modifications

Elimination of harmful interaction in the system is performed by introducing a third substance S_3, between substances S_1 and S_2, which is a substance S_1 and S_2, or modification (indicated S_1', S_2').

In this case, S_3 is introduced and is not foreign unlike model (4.1). System resources are used—existing system substances, S_1 or S_2, or their modifications S_1', S_2'.

This is described in model (4.4).

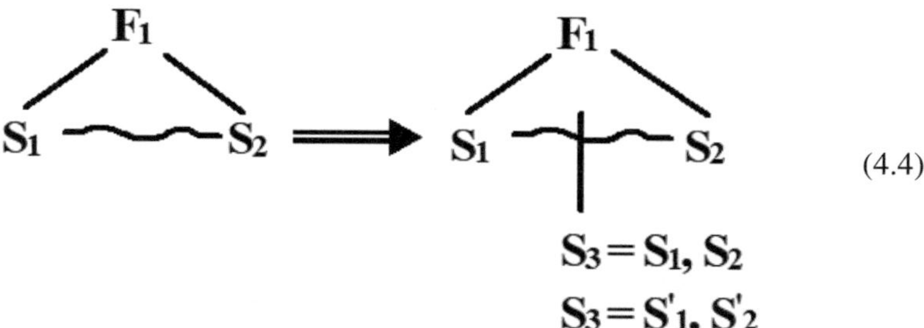

$$(4.4)$$

This is a more ideal model since we do not introduce additional substances but use only existing ones.

Problem 10. Warmer

Conditions of problem

A hot water bottle (Fig. 4.8) which has just been filled with boiling water can burn the patient.

What can be done?

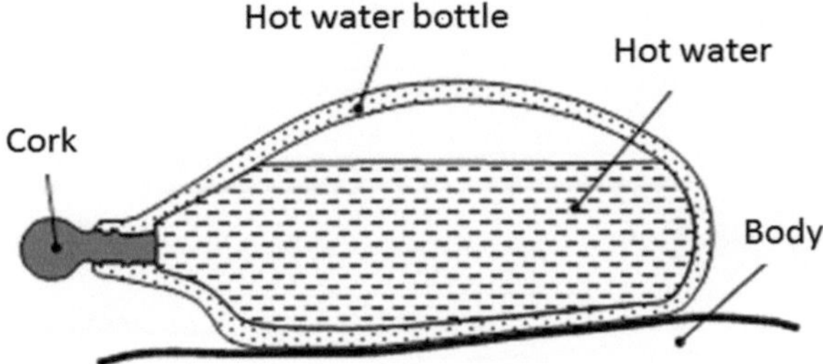

Fig. 4.8 Hot water bottle

Analysis of problem

The problem is in the form of Su-field type (4.5).

$$\mathbf{(S_1,\ S_2)} \rightarrow \mathbf{S_3} \qquad \mathbf{F_1} \tag{4.5}$$

where

S_1 water
S_2 hot water bottle
S_3 body of patient
F_1 thermal

Water S_1 is heated using thermal field F_1. Hot water S_1 heats hot water bottle S_2, and the latter warms (straight arrow) and can also burn (wavy arrow) the patient's body S_3. This is an Internal Complex Su-field with useful and harmful interactions.

Harmful interaction can be eliminated by introducing S_4, such as a towel that wraps around the hot water bottle, corresponding to model (4.1).

It is possible to present the solution as in model (4.4). S_4 can be represented as S_1, S_2, S_3 or modifications S'_1, S'_2, S'_3. Based on the solution in model (4.4), the structured solution can be represented in the form of (4.6).

$$\mathbf{(S_1, S_2)} \rightarrow \mathbf{S_3} \implies \mathbf{(S_1, S_2)} \rightarrow \mathbf{S_3} \qquad \mathbf{F_1} \tag{4.6}$$

$$S_4 = S_1, S_2, S_3$$
$$S_4 = S'_1, S'_2, S'_3$$

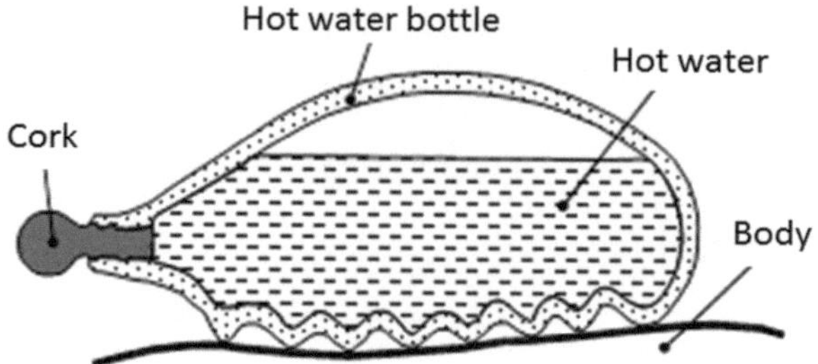

Fig. 4.9 Hot water bottle

There are ridges on one side of a German-patented hot water bottle (Fig. 4.9). Thus, the body is concerned only with the individual points of the hot water bottle and the layer of air between the body and the hot water bottle.

This invention uses a modification of a hot water bottle, $S_2{}'$.

We continue analysis of Problem 9 (hydrofoil).

According to model (4.4), S_3 can be used in the form of wing, water, or modification.

Examples of elimination of harmful interactions by using the substances themselves (wing and water) shall be demonstrated.

Example 15. Additional wing

In order to avoid cavitation, it is possible to add an additional wing S_3 (Fig. 4.10). This wing creates a flow that carries cavitation bubbles away from the wing. Thus, the wing is not damaged.

Example 16. Flow of water on wing

In order to avoid cavitation, it is possible to use water S_3.

1. The liquid flow on the wing can be created by making a thin wing with holes from end-to-end. (Fig. 4.11). Then, due to the pressure difference (P_1 and P_2), water from the lower part of the wing will be "sucked" to the upper surface of

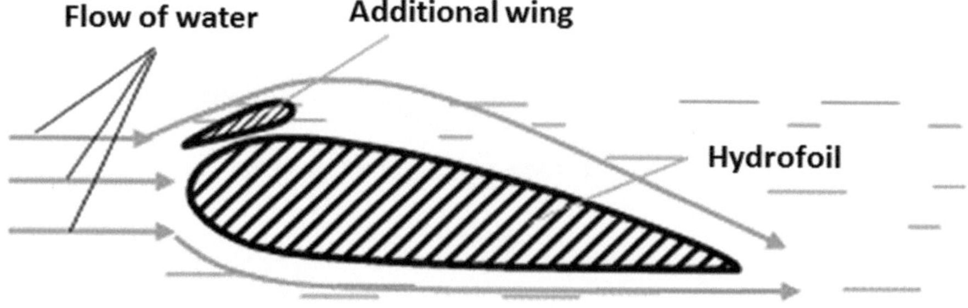

Fig. 4.10 Hydrofoil with an additional wing

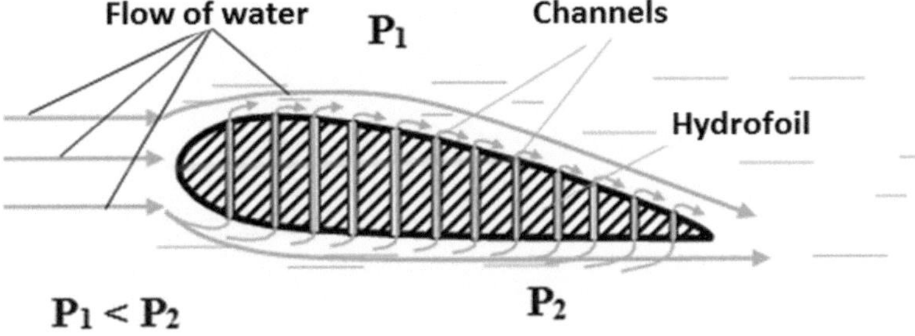

Fig. 4.11 Hydrofoil with additional flow of water

the wing. Recall that the difference in pressure over the wing and under the wing is generated by wing shape. The perimeter of the upper portion is longer than the bottom, so that the velocity of flow is higher at the top than at the bottom, in accordance with Bernoulli Principle where pressure is lower when the velocity of flow is higher.

Where:

P_1 pressure above wing
P_2 pressure below wing

2. There are grooves made in the wing (Fig. 4.12) that twist water flow, creating a layer of water on the wing surface to prevent cavitation. Furthermore, such a layer reduces water resistance which is a sort of "lubricant."
3. Suction of part of the liquid (boundary layer) adjacent to the body (suction improves velocity by approximately 1.5 times under constant force of installed energy). Suction can be arranged by moving incoming flow to the longitudinal channel in the center of the wing and transverse channels connected with the longitudinal bore at right angles (Fig. 4.13). Suction is created by a vacuum produced by fluid flow passing perpendicularly to the vertical channels. The phenomenon of injection is used.

Examples of eliminating harmful interactions by using modification of substances (wing and water) shall be demonstrated.

Fig. 4.12 Hydrofoil with grooves

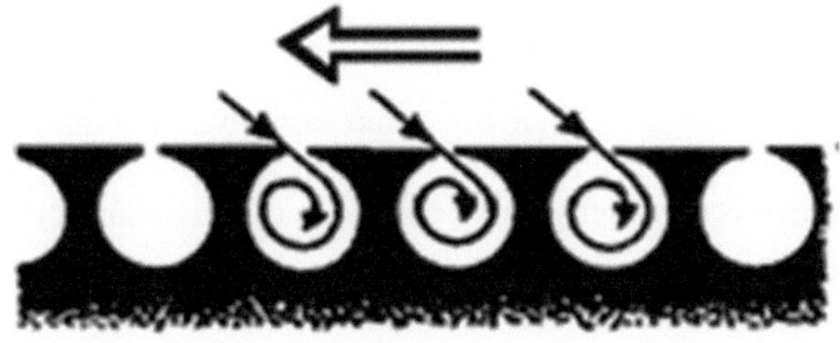

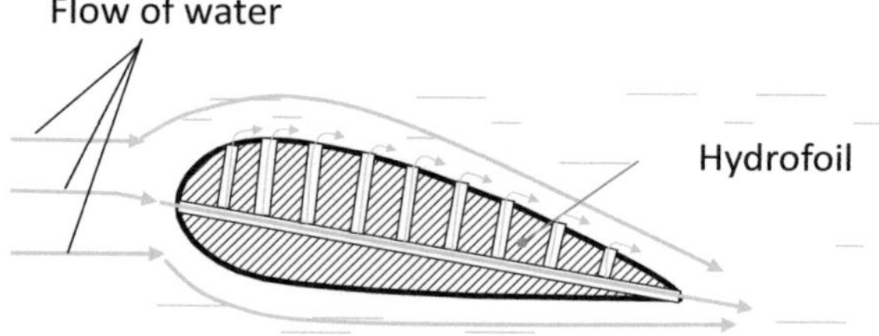

Fig. 4.13 Hydrofoil with additional flow of water

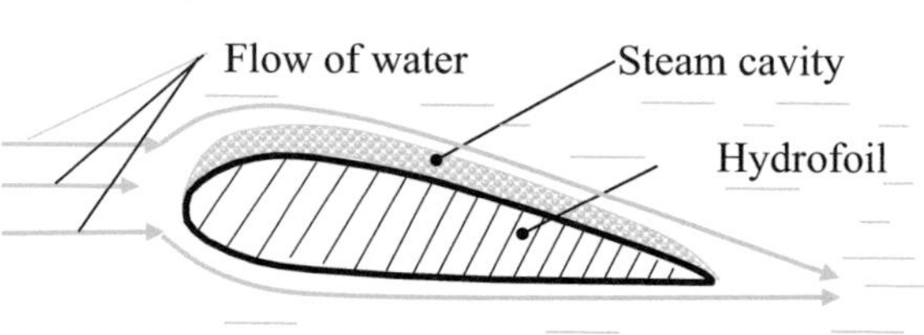

Fig. 4.14 Hydrofoil with steam cavity

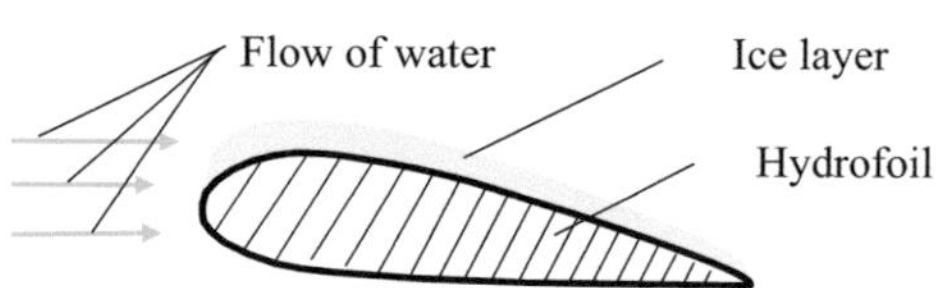

Fig. 4.15 Hydrofoil coated with ice layer

Example 17. Steam cavity

Cavitation can be prevented if the wing moves in a gaseous atmosphere—gas cavity (gas bubble around the wing). The gas cavity can be obtained by:

Modification of water

1. Convert water into a gas—steam. If the wing is heated, then a steam bubble (steam cavity) is formed. The cavity (Fig. 4.14) will not only prevent the wing from erosion, but also reduce resistance to movement of the wing in water.

   $S_3 = S_1'$ steam

2. Distribute water into oxygen and hydrogen.

   $S_3 = S_1'$ oxygen and hydrogen

Example 18. Ice layer

In order to prevent cavitation erosion, hydrodynamic shapes such as hydrofoils use a protective cover in the form of an ice layer itself (Inventor's certificate 412,062), which is constantly being frozen on the wing surface (Fig. 4.15).

$S_3 = S_2'$ ice

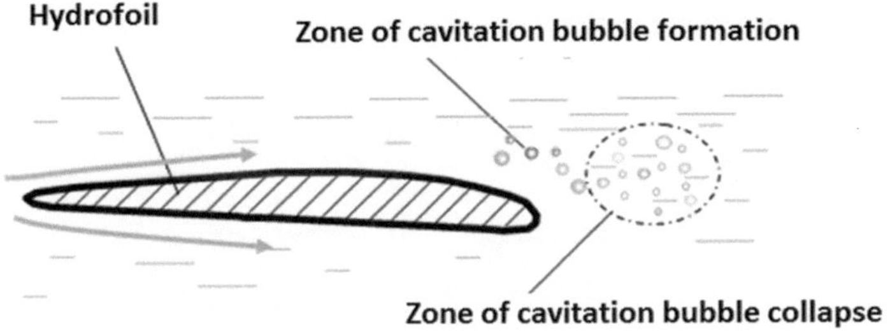

Fig. 4.16   Modification of hydrofoil

Fig. 4.17   Super-cavitation of hydrofoil creating air cavity

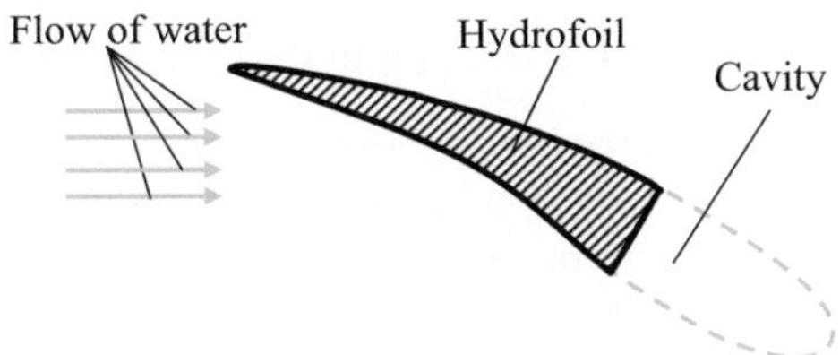

Example 19. Modification of wing

1. It is possible to change the wing shape, so that cavitation bubbles are formed only on the trailing edge of the wing and water flow is limited beyond it. Thus, cavitation of the bubbles will not occur on the wing. Modification of wing profile (shape) is a kind of turning itself by 180° (Fig. 4.16).
2. It is possible to change the wing shape, so that it creates an air cavity by itself. This is done by creating super-cavitation. The front part of the wing looks weird. There is a shockwave that splits the water flow such that it bypasses the wing (Fig. 4.17).

$S_3 = S'_2$ other wing profile

Strictly speaking, Figs. 4.11, 4.12 and 4.13 of Example 16 also demonstrate wing changes.

4.4   Elimination of Harmful Interactions by Introduction of Substance $S_3 = S'_1$, S'_2, and Field F_2

Modified substances, S'_1 or S'_2, can be available or be in place.

Elimination of harmful interactions in the system is performed by introducing modified substances of S_1 (S'_1) or S_2 (S'_2) between substances S_1 and S_2. This is achieved by introducing an additional field F_2 that acts on existing substances S_1 or S_2, modifying them into S'_1 or S'_2.

This can be represented in the form of model (4.7).

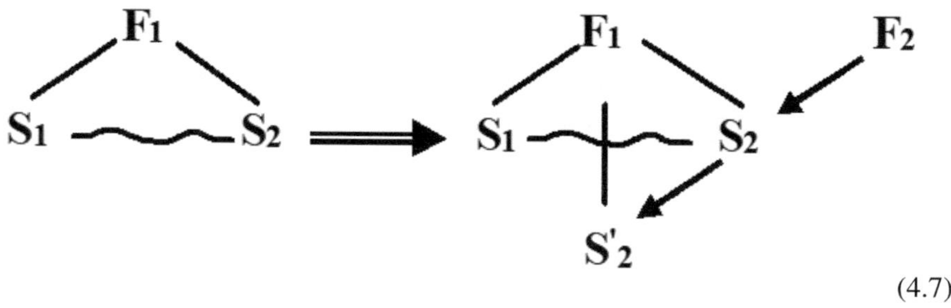

$$(4.7)$$

Example 20. Reducing hydrodynamic resistance

In order to reduce the hydrodynamic resistance of moving bodies such as vessels, an electromagnetic field generating complex molecules is created to reduce frictional forces at the boundary layer. In this invention, high-molecular compounds do not enter into the boundary layer, but instead a modified external environment, S_2', is used, through the action of an electromagnetic field. Furthermore, this invention can be used to reduce fluid resistance in the pipeline.

Figure 4.18 shows one of the variants described in Inventor's certificate 364 493. The nose section of the object which is moving in a fluid, is made of aluminum or iron. It is connected to the positive polarity of the current source, and the body is connected to the negative polarity. There is an insulating spacer between the body and nose section. When voltage is applied, aluminum hydroxide particles—$Al(OH)_3$, are formed at the boundary layer which reduces hydrodynamic resistance, similar to polymer additives being applied at the boundary layer. The surrounding environment is used when generating $Al(OH)_3$ particles.

In this solution, physicochemical effects are used.

Su-field model (4.7) has the form of (4.8) for this invention.

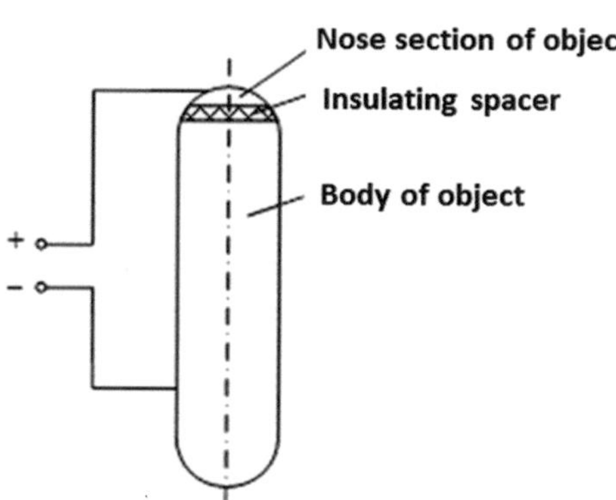

Fig. 4.18 Reducing hydrodynamic resistance—inventor's certificate 364,493

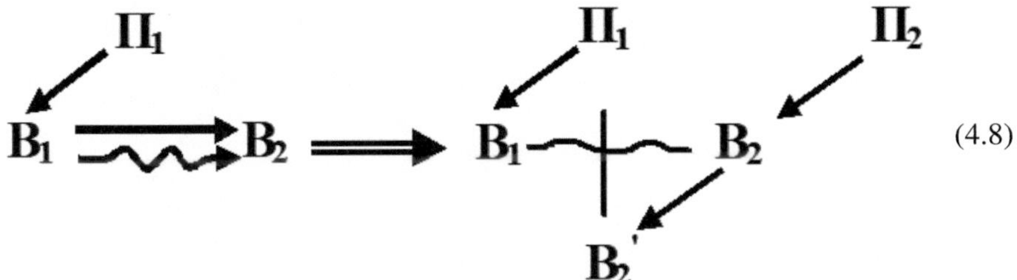

$$(4.8)$$

In this example: S_1—water, S_2—ship, hydrofoil, etc., F_1—flow of water, F_2—electromagnetic field, S_2'—complex molecules.

4.5   "Decelerate" Harmful Effects

Eliminating the harmful effects of field F_1 on substance S_1 is carried out by introducing a second substance S_2 to decelerate the harmful effects of field F_1.

Deceleration of harmful effects can be represented in the form (4.9).

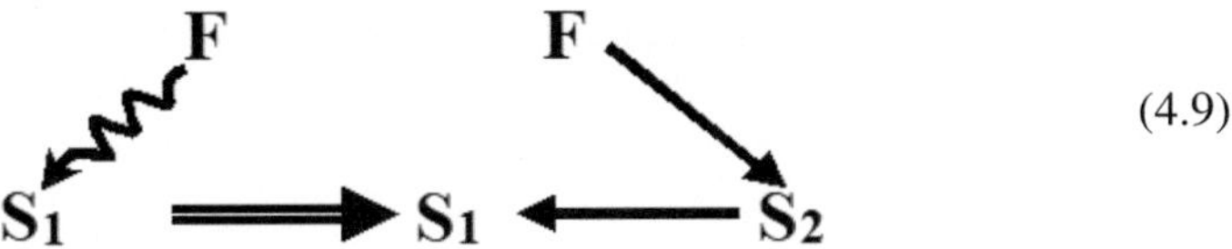

$$(4.9)$$

Example 21. Fuse

A network cable can burn out if there is a sharp increase in current. In order to avoid this, a fuse is used which can be either disposable (when a fuse is burnt) or reusable—automated.

4.6   Elimination of Harmful Interactions by Introducing F_2

The harmful effect is eliminated by transitioning to a double Su-field which neutralizes the harmful effects of field F_1 by field F_2.

This can be represented in (4.10).

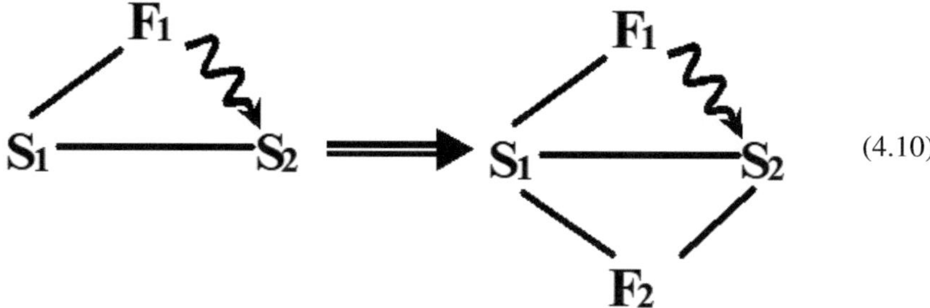

$$(4.10)$$

Problem 11. Artificial fireball

Conditions of problem

The artificial fireball (lightning ball) was researched under the guidance of academician P. L. Kapitsa in the laboratory. There was a sealed cylindrical quartz chamber filled with helium at a pressure of 3atm (Fig. 4.19). During the helium plasma discharge and under the influence of a powerful electromagnetic field, the plasma is pulled together into a spherical plasmoid—"fireball." In order to keep the fireball in the center of the chamber, a solenoid ring is used which is located around the chamber. Based on experimental research, it was necessary to increase the electromagnetic radiation power in order to increase the power of the fireball.

The plasma became hotter, and therefore denser. The fireball became lighter and rose up, touching the chamber walls and damaging them. Electromagnetic forces did not balance buoyancy forces. In order to keep the fireball in the center of the chamber, the power of the magnetic field from the solenoid was increased, but nothing happened: The fireball rose up—just a little slower. The staff offered to build a new installation with a more powerful solenoid, but P. Kapitsa acted differently. How?

Analysis of problem

Formulate the problem in the form of the Su-field model (4.11).

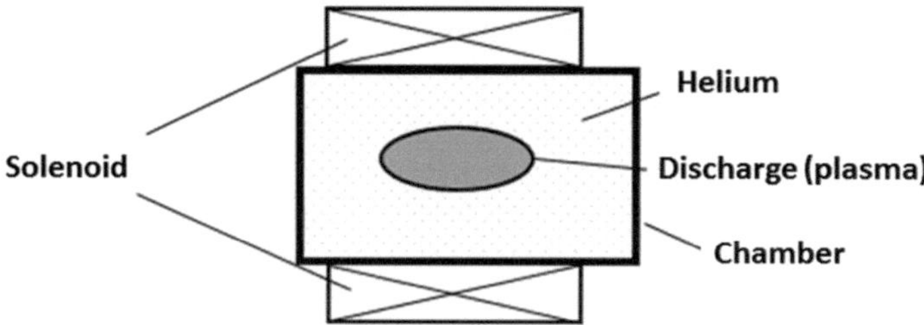

Fig. 4.19 Apparatus for the creation of artificial fireball

$$(4.11)$$

Given an inefficiently controlled Su-field:

S_1 lightning
F_1 gravitational field acting on fireball
S_2 gas that does not balance the action of the gravitational field

In order to improve the controllability of the particular Su-field, it is necessary to counteract field F_2 according to model (4.12).

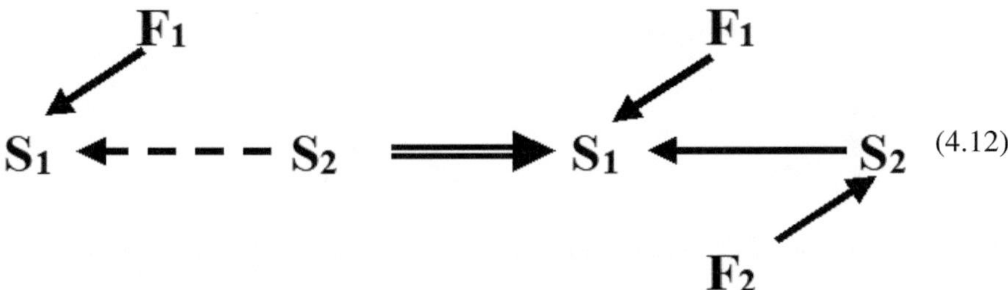

$$(4.12)$$

The field F_2 must counteract the gravitational field F_1. The most effective would be to use an electromagnetic field but it would require installation to be redone. In line with the trend of Su-field evolution, the mechanical field is the initial field to be used. The most effective in this case–the centrifugal force field.

P. Kapitsa proposed to spin the gas, giving it a continuous rotation. As the gas was spun, it ceased to discharge to the surface. The gas was forced to continuously revolve around the blower, which is similar to a home vacuum cleaner. So, the home vacuum cleaner concept was used in the early stages (Fig. 4.20).

F_2 centrifugal field

4.7 Elimination of Harmful Interactions by Introducing S_3 and F_2

Introduce substance S_3 which generates field F_2 to neutralize the harmful effects of field F_1, which eliminates the need to transition to a mixed Su-field (4.13).

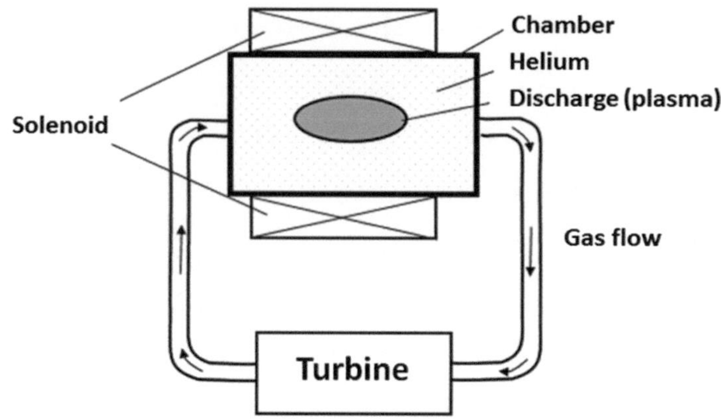

Fig. 4.20 Creation of centrifugal forces by means of vacuum cleaner

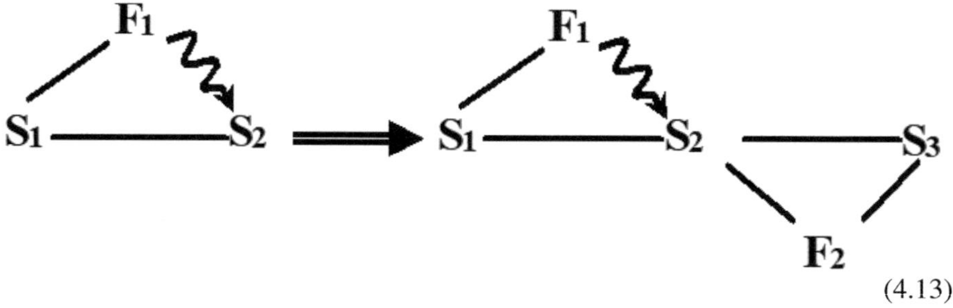

$$(4.13)$$

Problem 11. Artificial fireball (continued)

Field of centrifugal forces F_2, is created by turbine S_3 (4.14).

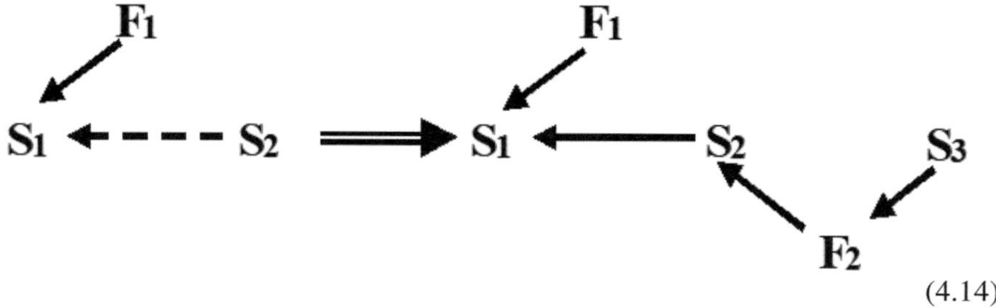

$$(4.14)$$

4.8   Elimination of Harmful Interactions by Introducing S_3, F_2, and F_3

The harmful effect is eliminated by transitioning to a mixed Su-field in which the introduced substance S_3 influences field F_3 to generate field F_2, in order to neutralize the harmful effects of field F_1 (4.15).

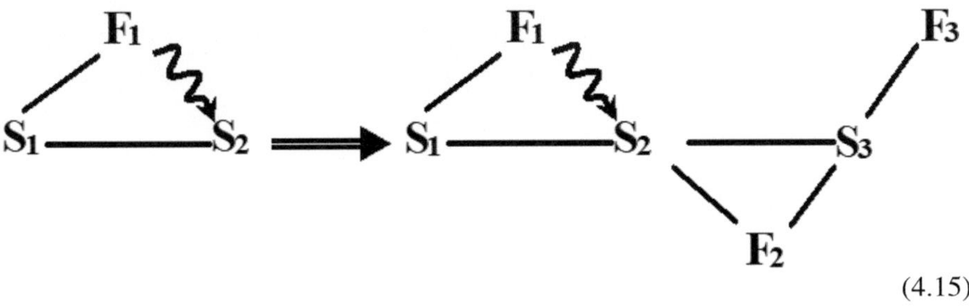

$$(4.15)$$

Problem 12. Ball extraction

Conditions of problem

It is not uncommon when you need to extract a ball that is seamed to a body (Fig. 4.21). In order to do this, we have to break the seam.

How to extract the ball without breaking the seam?

Analysis of problem

Formulate the problem in the form of the Su-field model (4.16).

$$(4.16)$$

The Su-field has a harmful interaction:

Fig. 4.21 Ball seamed to body

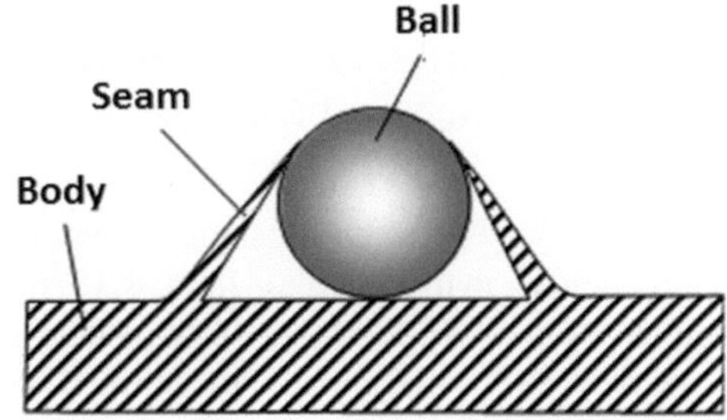

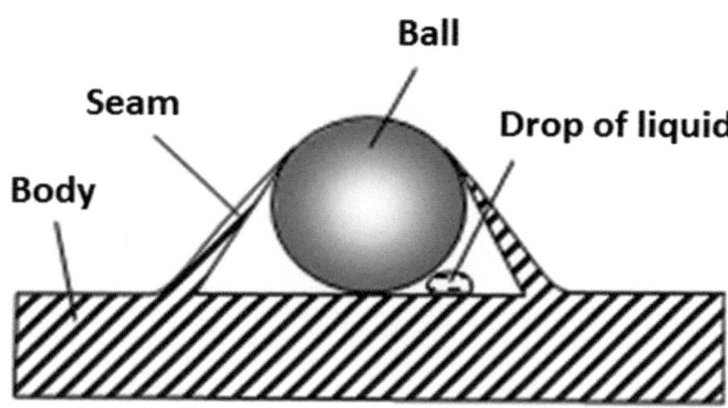

Fig. 4.22 Introduction of liquid drop beneath the ball

S_1 body
F_1 mechanical field holding the ball in the body
S_2 ball

Harmful interaction can be eliminated by introducing S_3 which generates field F_2, which neutralizes harmful effects of field F_1. In accordance with model (4.15) for this problem, a structural solution can be represented by model (4.17).

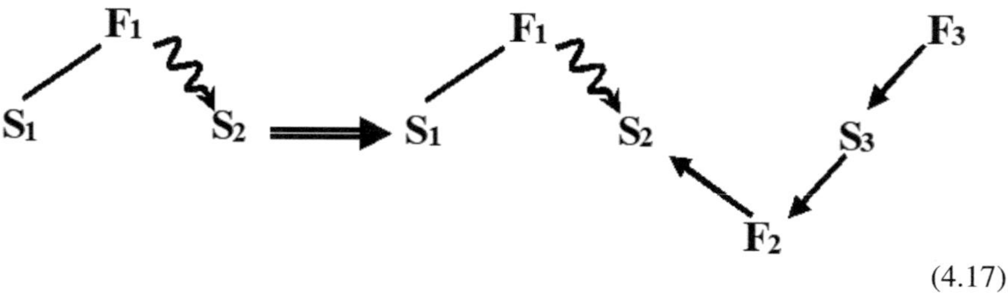

$$(4.17)$$

It is necessary to introduce S_3 which creates F_2 which pushes the ball.

One of the solutions.

Inject a drop of liquid S_3 beneath the ball S_2 (Fig. 4.22). If necessary, S_3 is heated (field F_3) to evaporate the liquid that creates pressure (field F_2) to push ball S_2 from body S_1.

4.9 Elimination of Harmful Interactions Between Substance and Field by Introducing S_2 and F_3

Harmful effects between substance and field are eliminated by introducing substance S_2 and field F_3 in modified Su-field (4.18) or introduction of substance S_2 and field F_2 in measurement Su-field (4.19), or instead of field F_2 add a third modification of the same field F_1''' (4.20). The incorporated field (F_3, F_2, or F_1''') acts upon substance S_2, changes feature of S_2, controls field F_1''.

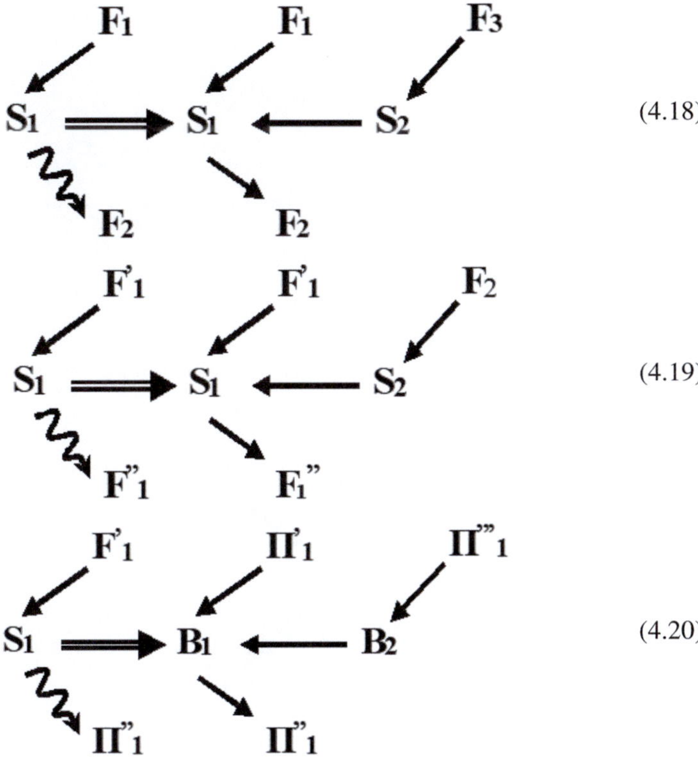

$$(4.18)$$

$$(4.19)$$

$$(4.20)$$

Problem 13. Automotive Glass

Conditions of problem

Driver can be blinded by lights from headlamps of the car behind him through a rear view mirror Fig. 4.23.

What can be done?

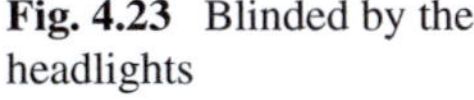

Fig. 4.23 Blinded by the headlights

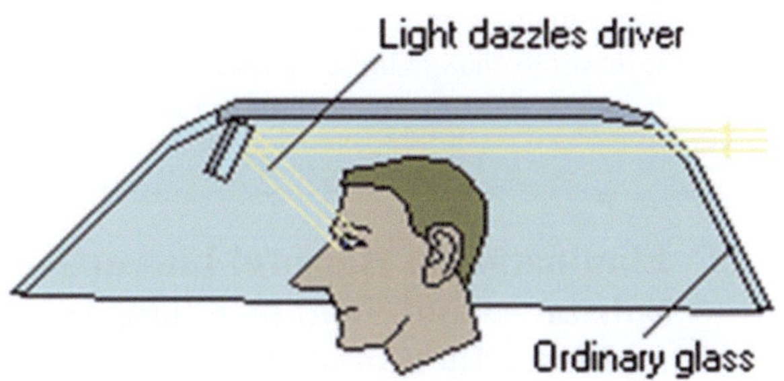

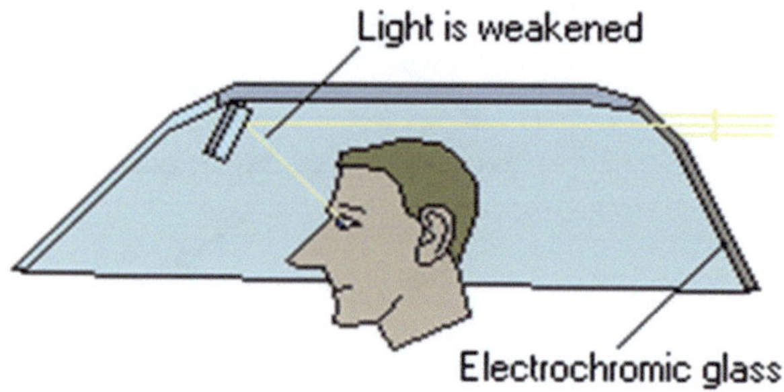

Fig. 4.24 Electrochromic film

Analysis of problem

Present problem in the form of Su-field (4.21).

$$\text{(4.21)}$$

Given a Su-field with harmful interaction:

S_1 rear-view mirror
F_1' lights from the headlamp of the car from behind (optical field)
F_1'' reflected light acting on the driver

Use model (4.19).

Analysis of problem

The defect can be eliminated if the rear window of the car to is covered with an electrochromic film S_2 (nickel oxide and conductive tin oxide layer). The transparency ratio of glass changes under electric current F_2. It is a controlled tinted glass. This allows the driver to change light intensity F_1'' entering from outside. Thus, controlled field F_1'' changes field F_2. Such films are used instead of blinds (Fig. 4.24).

4.10 Elimination of Harmful Interactions Between Substance and Field by Changing S_1 on S_2 and Introduction of F_3

Harmful effects between substance and field are eliminated by replacing substance S_1 with another substance S_2, and introduction of a field:

- F_3 (4.22)
- modification of F_2 in Su-field (4.23)
- third variation of field F_1''' (4.24)

Incorporated field (F_3, F_2, or F_1''') acts on S_2, which controls field F_2 or F_1'''.

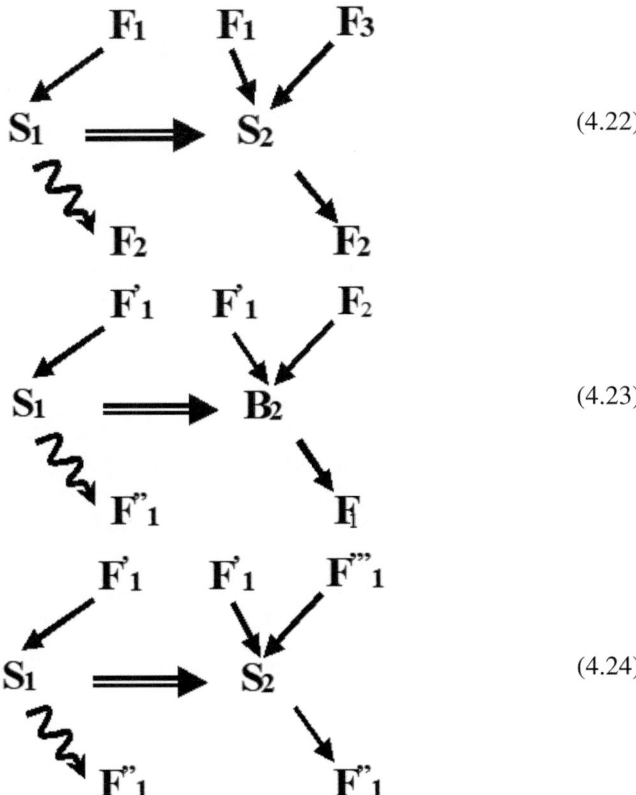

(4.22)

(4.23)

(4.24)

Problem 14. Diode

Conditions of problem

Diode passes current in one direction and does not pass in another. This is always consistent. Incorporate, for example, relays in order to enable a controlled process. This greatly complicates the system.

What can be done?

Analysis of problem

Formulate a problem in the form of Su-field (4.25).

$$(4.25)$$

Given Su-field with uncontrolled interaction:

S_1 diode
F_1' AC voltage
F_1'' DC voltage

The dotted arrow indicates that the action is insufficient (uncontrolled).

Use Model (4.24). Replace S_2 with S_1 in which it is possible to control output field F_1".

Solution of problem

Disadvantages can be eliminated by replacing diode S_1 with thyristor S_2. The thyristor performs the same function as the diode (passes current in one direction and does not pass in the opposite direction), but it has another control input. If the control input is not supplied with current F_1'", the thyristor will not pass current even in the forward direction. However, it is necessary to apply at least a short pulse as it immediately conducts and remains conductive as long as there is a direct voltage. If the voltage is removed or changed in polarity, the thyristor does not conduct.

Field F_1'"   current at control input (electric field)

Chapter 5
Finding the Desired Effect

Technological effect types (physical, chemical, biological, and mathematical, in particular geometrical) used in Su-field are defined as follows.

If substance S_1 converts one field F_1 to another field F_2, described by model (3.3), or change in parameters of field F' on F'', described by model (3.3), the desired technological effect is obtained by combining fields (5.1) and (5.2).

$$S_1 \begin{array}{c} F_1 \\ \\ F_2 \end{array} \Bigg\} \text{ Technological effect} \tag{5.1}$$

$$S_1 \begin{array}{c} F' \\ \\ F'' \end{array} \Bigg\} \text{ Technological effect} \tag{5.2}$$

In accordance with this, not only is the structure of the future solution being defined, but also the type of technological effect to be used, i.e., Su-field analysis is a tool for finding relevant technological effects (physical, chemical, biological, or mathematical) in providing solutions to specific problems. The final search for the desired effect is by using the effects index.

Here is an example of the use of model (5.1).

Example 22. Microphone

Microphone S_1 converts sound vibrations (acoustic field) F_1 (F_{acous}) into electrical F_2 (F_{electr}).

V. Petrov, *Structural System Analysis*, https://doi.org/10.1007/978-3-031-55825-2_5

The name of the desired effect—acousto-electric, according to the physical effects index to find suitable effects—piezoelectric or ferroelectric effects (5.3).

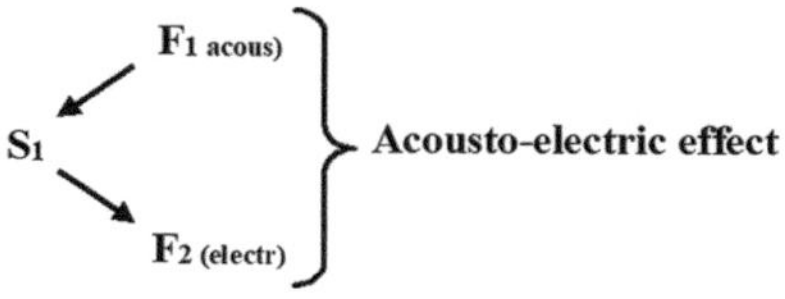

$$S_1 \left\langle \begin{array}{l} F_{1\ (acous)} \\ \\ F_{2\ (electr)} \end{array} \right\} \text{Acousto-electric effect} \qquad (5.3)$$

Here is an example of the use of model (5.2).

Example 23. Sonar

Ultrasound studies (US) or sonar work is as follows. Ultrasonic (acoustic) signal F' is transmitted and is reflected from object S_1. The reflected ultrasonic (acoustic) signal F" contains information about the object under study.

Where

S_1 object under study.
F' transmitted signal (acoustic field).
F'' reflected signal (acoustic field).

This similar principle (reflected signal) operates like a radar.

Chapter 6
Law of Increasing Degree of Su-Field

6.1 Introduction

The law of increasing degree of Su-field states that any system which is evolving tries to become more of a Su-field, i.e., must increase the degree of Su-field.

Initially, this law was developed by G. Altshuller.

Development of technical systems is going towards increasing the degree of Su-field. The meaning of this law is that non-Su-field systems tend to become Su-filed and in Su-field systems a development is going towards the transition from mechanical to electromagnetic fields; increase of degree of fragmentation and the number of connection of elements and system response [1].

In works [2, 3], Altshuller has described a mechanism of the law of increasing the degree of Su-field. A development is going from non-Su-field system to a simple Su-filed system, then to complex Su-filed (internal, external, and based on environment and modified substance of environment), then towards chain Su-field, double Su-field, and forced Su-field.

The following presents an improved version of the law of increasing the degree of Su-field by the author.

6.2 The Law of Increasing Degree of Su-Field

The law of increasing degree of Su-field states that any technical system in its development tends to become *more Su-field like*, i.e., its *degree of Su-field* has to increase.

The law includes trends, describing the sequence of changes in the Su-field model and components (substances and fields) in order to achieve a more controlled technical system, i.e., more ideal system. It is necessary to implement the coordination of substances, fields, and structures in the process of change.

V. Petrov, *Structural System Analysis*, https://doi.org/10.1007/978-3-031-55825-2_6

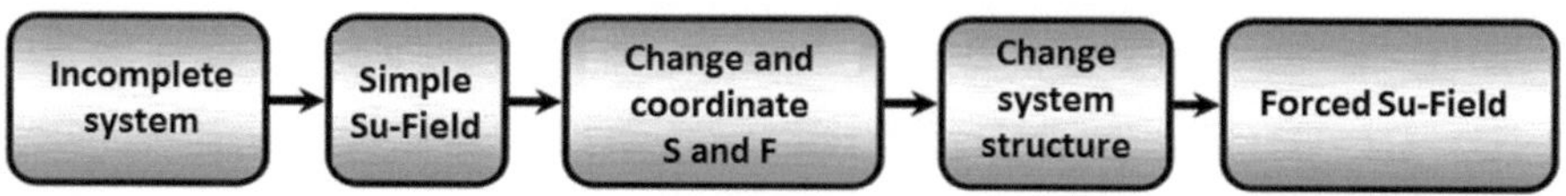

Fig. 6.1 General trend of Su-field development

The general trend of Su-field evolution is shown in Fig. 4.22. It is a transition: from a **non-Su-field** system (or an **Incomplete Su-field system**) to a **simple Su-field system**; next is a change and subsequent **coordination of substances (S) and fields (F)**; then **change Su-field structure**; and, finally, transition to **forced Su-field**.

Forced—greatly controlled Su-field.

Thus, Su-field evolution trends can be identified as a trend in building Su-fields. Other trends of Su-field analysis consider Su-field transformation in order to increase the efficiency of technical systems or elimination of harmful interactions. They are the consequence of the law of increasing degree of Su-field for technical systems. Su-field transformation can change components (substances and fields) and structure. These changes can be implemented partially or fully in space, in time, or on condition.

The general trend is shown in Figs. 6.1, 6.2, 6.3, 6.4, and 6.5.

The first trend in Su-field development—Complete (constructed) Su-field, i.e., transition from Incomplete Su-field to Su-field system was discussed above (Sect. 3.3) as a basic rule of Su-field analysis and described by Model (3.15). The result is a Simple Su-field.

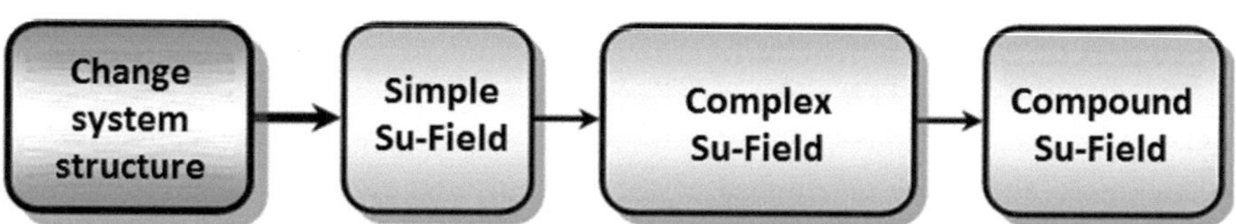

Fig. 6.2 Trend in the structure of Su-field

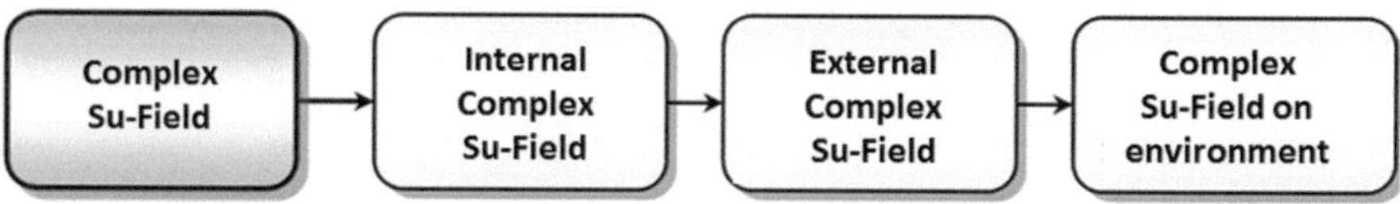

Fig. 6.3 Trend in complex Su-field

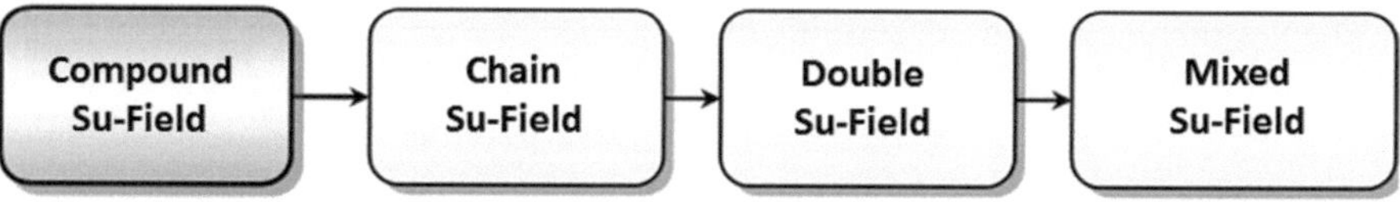

Fig. 6.4 Trend in compound Su-field

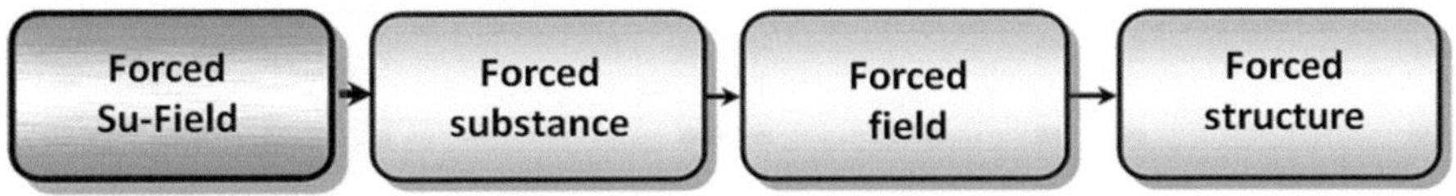

Fig. 6.5 Trend of forced Su-field changes

Change in substance (S) and field (F) begins with the selection of substances that are "responsive" to an existing field or fields that are "responsive" to an existing substance or "responsive" (substance-field) pair. Selected "responsive" substances and fields are coordinated.

The concept of "responsiveness" was discussed in Chap. 1.

In practice, it is advisable to immediately select the "responsive" substances and fields in the construction of the Su-field.

Change of substances and fields in Su-field is done to improve the functionality of the system. The replacement of a selected responsive substance and field provides better performance in the execution of the main function of the system.

Sometimes changes in substances and fields are sufficient to increase system efficiency.

Further system development is by changing its structure and use of Forced Su-fields. Coordination is necessary to be done after each change.

Trend of change in the Su-field model is shown in Fig. 6.1 and represents a transition from Simple to Complex Su-field and from Complex to Compound Su-field. This is done primarily by increasing the number of interactions between components and their quantity.

The next stage is the trend of development of **Complex Su-field** as shown in Fig. 6.2, i.e. transition from an Internal Complex Su-field (3.19) and (3.20), to an External Complex Su-field (3.23) and (3.24), followed by Complex Su-field on environments (3.28) and (3.29). Details are in Sect. 3.3.

This trend is driven primarily by the fact that additives are far easier to be introduced not into the system, but introduced from outside or easier to be carried by the environment. In addition, such additives are easy to be removed or replaced when necessary.

The trend of development of **Compound Su-field** (Fig. 6.3) represents a transition from Chain Su-field (3.38) to Double (3.42), and Mixed Su-field (3.45) and (3.46).

The highest stage of increasing the controllability of Su-fields is a transition to Forced Su-field. The trend of development of **Forced Su-field** is shown in Fig. 6.4.

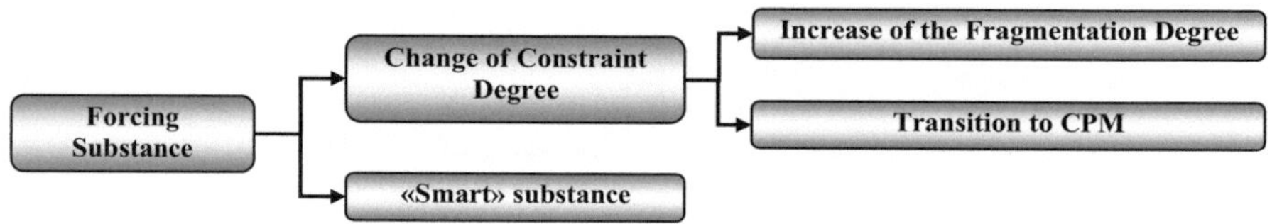

Fig. 6.6 Trend of substances forcing

6.3 Forced Su-Field

6.3.1 General Trend

Forced Su-field—Su-field using controlled substance, field, and structure.
Details of this trend are not covered in this book. It can be found in [4].[1]

6.3.2 The Forcing of Substances

The forcing of substances is done by the transition toward **more controllable substances** that are subject to the Change of Constraint Degree Law and by the transition to more *progressive* **("smart") substances**.

This trend is shown in Fig. 6.6.

The Change of Constraint Degree Law is substantiated by trends of **Degree of Fragmental Increase** [5] and by **transition to the capillary-porous materials (CPM)** [6]. This trend is shown in Fig. 6.7.

"Smart" substance—this material is "responsive" to a certain field. It is capable of carrying out a specific function through the use of effects (physical, chemical, or biological) under the influence of a field.

Examples are liquid crystal, polarizing plate, substances that change transparencies; thermal and photosensitive polymers, fluorescent material, polymeric gels, shape memory material, magnets, magnetic rheological liquid, electrets, heat pipes.

"Smart" substance can also be defined as a converter or source, to carry out certain effects (physical, chemical, biological, or geometrical).

---

[1] Petrov Vladimir. The laws of system development: TRIZ. Ed. 2nd, rev. and supplemented/Vladimir Petrov. Publishing Solutions, 2019.—922 c.—ISBN 978-5-4490-9985-3 (Russian) https://ridero.ru/books/zakony_razvitiya_sistem.

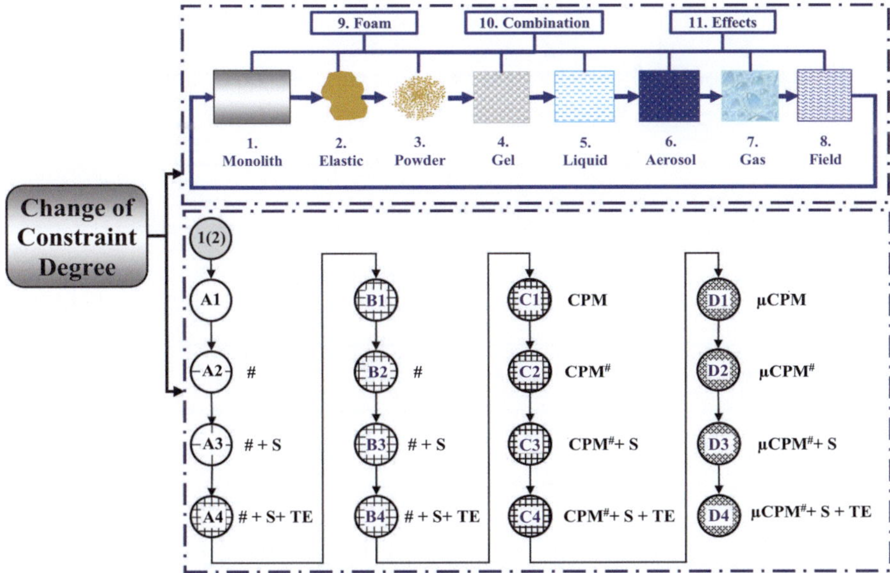

Fig. 6.7   Trend of change of constraint degree law

6.3.3   *Forcing of Fields*

The forcing of Fields is a subject of **Transition to Micro-level Law** and **Increase of Energy Density and Information Density law** [7]. The trend of forcing Fields is shown in Fig. 6.8. As a rule, the usage of more controllable fields is associated with the use of technological effects.

The main trend of changes of a field type to a more controlled field is shown in Fig. 6.9.

Consider the chemical field in other areas of technology.

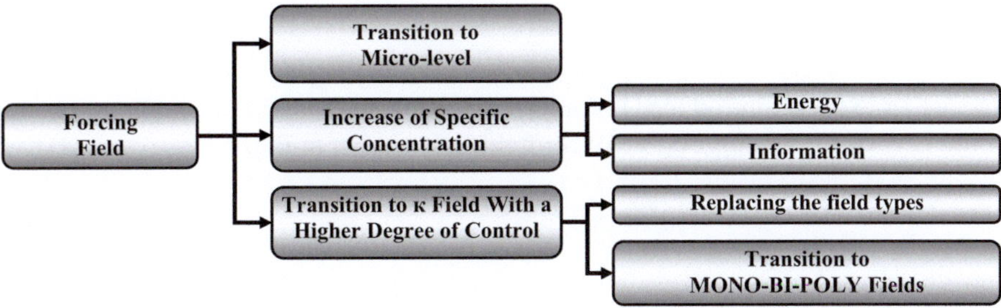

Fig. 6.8   Trend of forcing fields

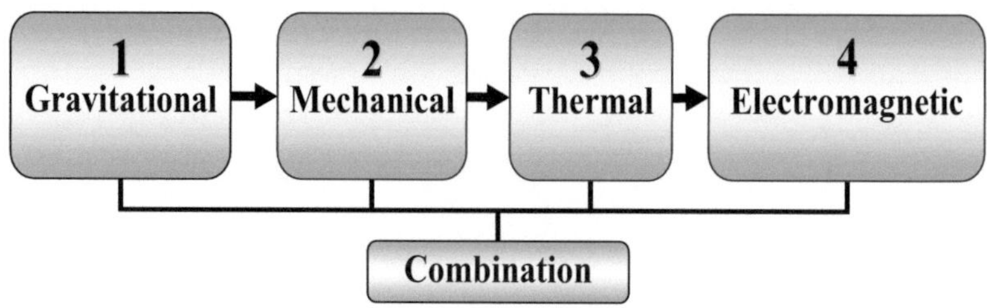

Fig. 6.9 Trend of changes of a field type to a more controlled field

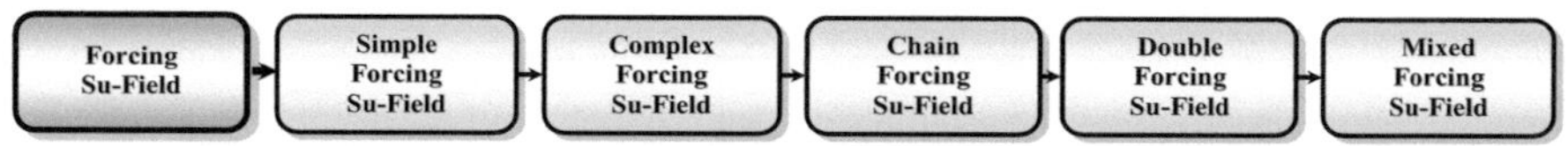

Fig. 6.10 Trend of forcing structure of Su-field

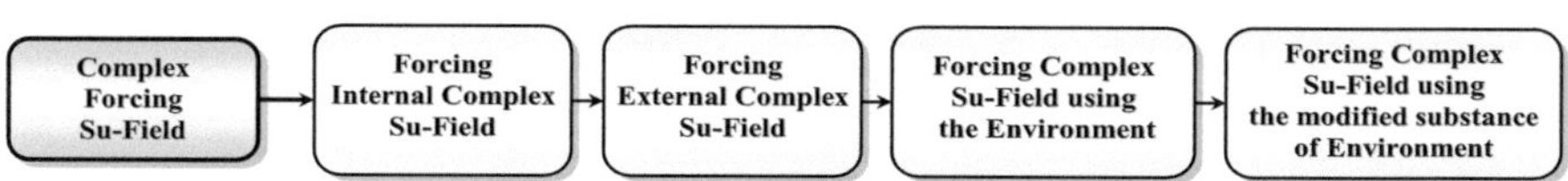

Fig. 6.11 Trend of evolution of complex forcing Su-field

The forcing of the Su-field structure is done by the transition from simple forcing Su-field to complex forcing Su-filed, then to chain forcing Su-filed, then to double forcing Su-filed and mixed forcing Su-field (Fig. 6.10).

Complex forcing Su-fields (Fig. 6.11) can be internal and external complex forcing Su-fields, and complex forcing Su-fields using an environment (using a substance of environment, its modifications or addition into the environment).

6.4   The Detailed Scheme of the Law of Increasing Degree of Su-Field

According to trends introduced before, it is possible to compose a complete scheme of the Law of Increasing Degree of Su-field. It is shown in Fig. 6.12.

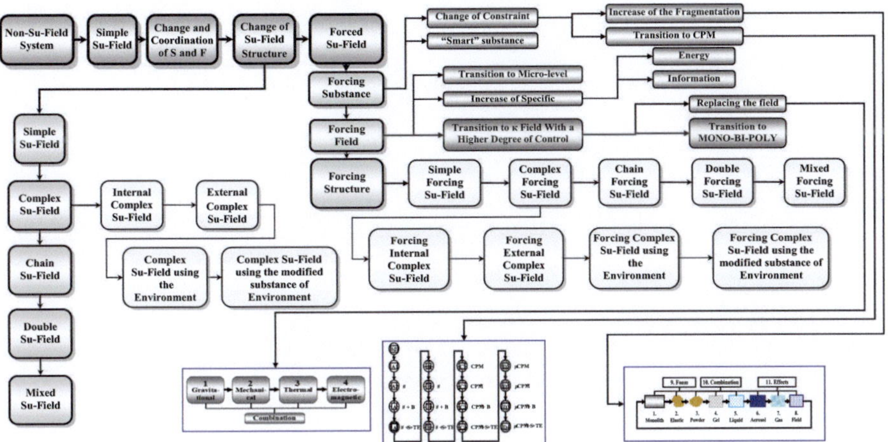

Fig. 6.12 The scheme of the law of increasing degree of Su-field

Chapter 7
Su-Field Analysis for Information Systems

In information systems and especially in programming, there are no substances and fields.

Concerning Su-field analysis was renamed **El-Action**, where **El** (**Element**) stands for *Substance* and **Action** for *Field*.

The new name is more suitable for the analysis of information systems. It can be used in the analysis of all technical systems as well.

In Figs. 7.1, 7.2, and 7.3, the El-Action increasing degree law is shown (Fig. 7.4).

Chapters 7 to 8 are a joint work with German Voronov.

Petrov V., Voronov G. A New Approach to Su-Field (structural) Analysis/Further development of Su-Field Analysis. Development of Inventive Thinking./Collection of Scientific Papers. TRIZ Developers Summit Library. Issue 5. Kiev, 2013.—258 pages, P. 166–188.

Vladimir Petrov, German Voronov. A New Approach to Su-Field Analysis: TRIZ Kindle Edition. Amazon Digital Services LLC. ASIN: B01L8SU5XO.2016.—39 p.

© The Author(s), under exclusive license to Springer Nature Switzerland AG 2024

V. Petrov, *Structural System Analysis*, https://doi.org/10.1007/978-3-031-55825-2_7

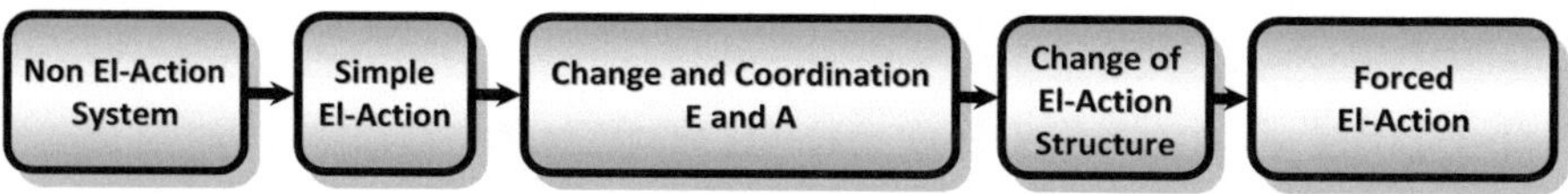

Fig. 7.1   General trend of El-Action

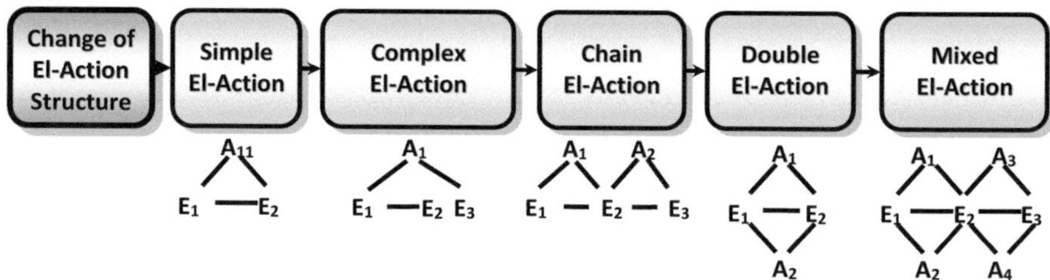

Fig. 7.2   El-Action structure change trend

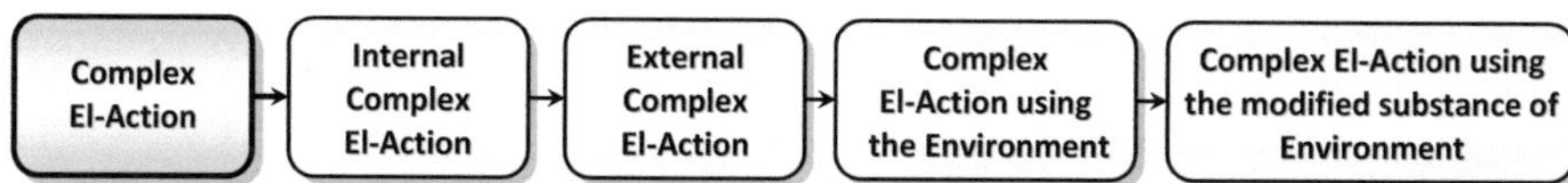

Fig. 7.3   Complex El-Action change trend

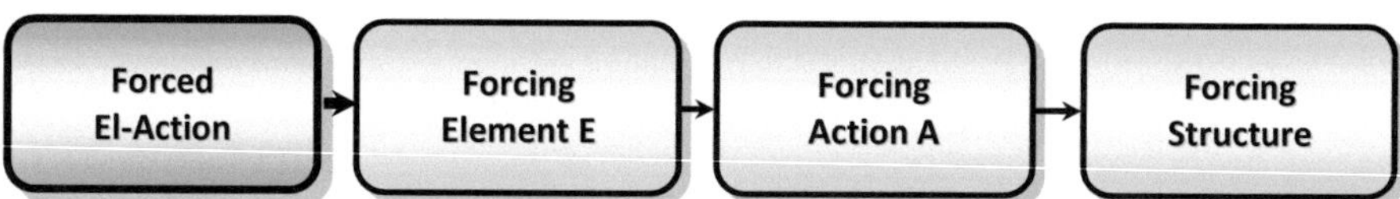

Fig. 7.4   Forced El-Action change trend

Chapter 8
The New Su-Field Structure

8.1   General Concepts

In this book, we introduce a new Su-field (El-Action) structure.

The Su-field contains two components: Substance and Field and the **El-Action** contains **Element (E)** and **Action (A)**. We have added a third component **Knowledge (K)**.

The new model, which includes Element (E), Action (A), and Knowledge (K) is called **EAK**.

The analysis method and transformation of EAK will be called **EAK Analysis**.

Knowledge (K) development in the system has four states:

1. Knowledge is outside the system (Fig. 8.1).
2. Partial knowledge is included in the system during the design stage, while the remaining knowledge is placed into the super-system (Fig. 8.2).
3. All the necessary knowledge is contained within the system, but the knowledge control is outside (Fig. 8.3) of the system (in the super-system).
4. Knowledge control is performed inside the system (Fig. 8.4).

Example 24. Need to drill a hole

1. Knowledge outside the system

Drilled by hand. Action (A)—rotation. It acts on the Element (E)—drill. Knowledge (K) is located outside the system. The driller knows where to drill the hole and how to do it.

$$A \rightarrow E \tag{8.1}$$

V. Petrov, *Structural System Analysis*, https://doi.org/10.1007/978-3-031-55825-2_8

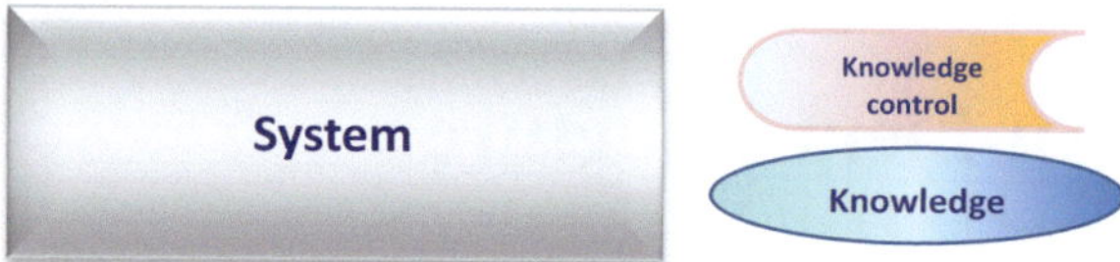

Fig. 8.1 Knowledge is outside the system

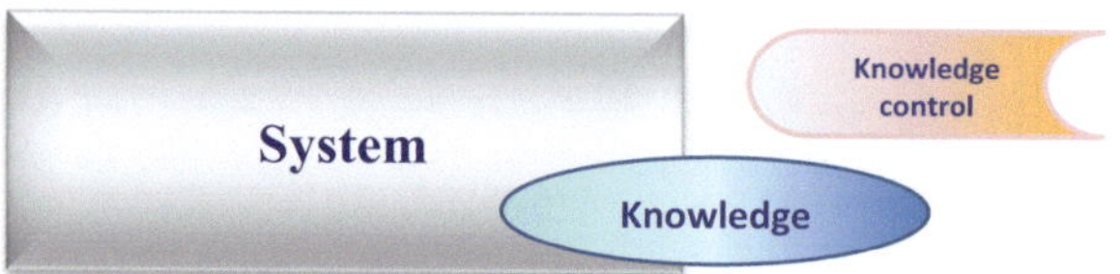

Fig. 8.2 Partial knowledge

Fig. 8.3 Knowledge is contained within the system

Fig. 8.4 Knowledge is
performed inside the system

where

A   Action
E   Element

2.   Partial knowledge included in the system

Special drilling jig is used for drilling. It holds the drill at the location, where the hole
should be made. The driller is not responsible for the location and punch. Knowledge
about these operations is available in the jig (Fig. 8.5).

   Knowledge (K) controls the Action (A) that acts on an Element (E). Knowledge
of how to make a hole is outside the system (in the operator's mind). The dotted
arrow (8.2) indicates that we used partial knowledge.

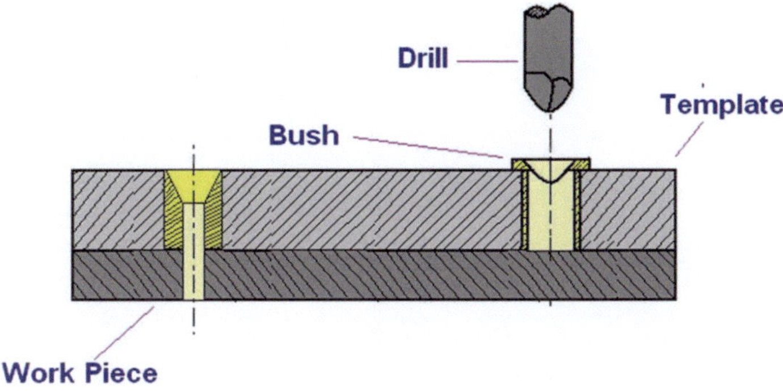

Fig. 8.5 Jig

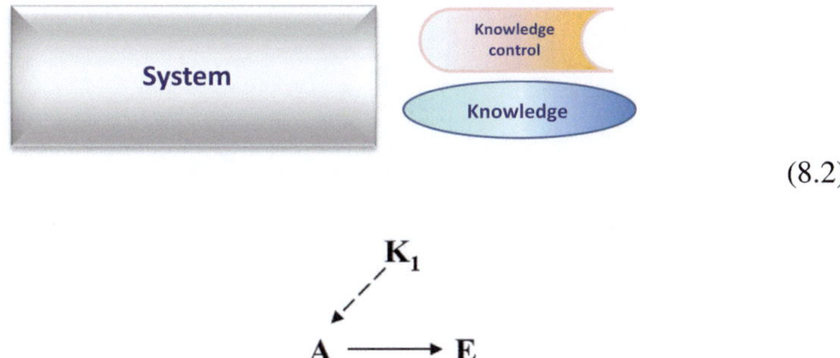

$$(8.2)$$

where

K   Knowledge
A   Action
E   Element

3. All knowledge of the process is in the system

A CNC machine has all the necessary knowledge to implement technology manu-facturing. Control of this knowledge (programming) is outside the system in the programmer's mind.

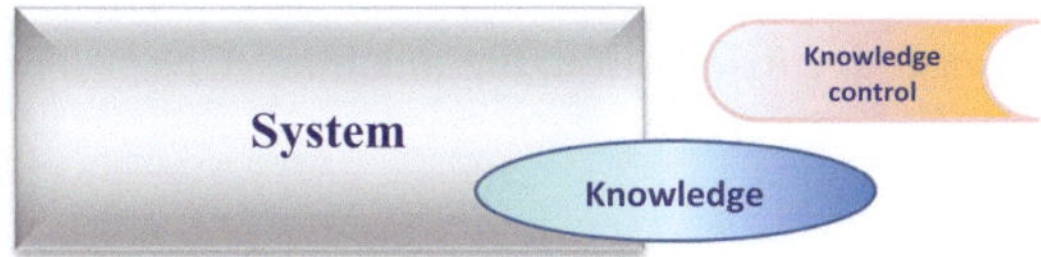

   Control of this knowledge (programming) is outside the system in the programmer's mind (Fig. 8.6).

Fig. 8.6 CNC—computer numerical control

$$A \xrightarrow[\searrow{K_1}]{} E \tag{8.3}$$

where

K   Knowledge
A   Action
E   Element

4.   Knowledge control is performed inside the system

Programming (K_2) must be carried out in the machine (CNC). This is the next stage of development.

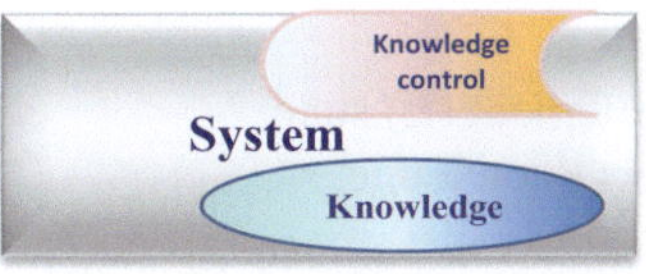

$$(8.4)$$

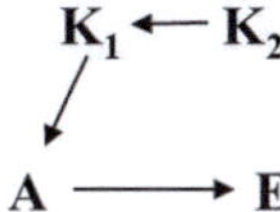

where

K Knowledge

- **K_1** knowledge, control action
- **K_2** knowledge, control knowledge

A Action
E Element

In general, states 2–4 may be represented by more complex models.

An Element (E) may initially contain some Knowledge (K), for example, the element status. In order to control element (E), it is often necessary to know its status. Then model (8.2) can be written as:

$$(8.5)$$

In state 3, when control is needed for element (E), model (8.3) can be written as follows:

$$(8.6)$$

This case is typical for any self-adjusting system, such as a homing missile.

Controlling the state of an Element (E) for state 4, can be represented as:

$$(8.7)$$

where

K₁   knowledge, control action
K₂   the knowledge of the Element (E) state
K₃   knowledge that controls knowledge

In the homing missile example, K_3 may be thought of as a change of purpose, the abolition of an action or self-destruction, etc.

Accounting knowledge and development patterns are modern trends in the technology. In Information Technology, developing the accounting knowledge is of the utmost importance.

Often the Su-fields are represented as a triangle. The same thing can be done for the El-Action. Therefore, in the general case, EAK can be represented as follows:

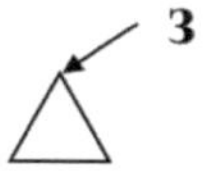

$$\text{(8.8)}$$

Thus, Su-field analysis is a special case of EAK analysis, provided knowledge is ignored or not considered in the system analysis and synthesis.

To complete the picture, it is necessary to take into account changes in the elements (E), action (A) and knowledge (K) in time, i.e., their dynamic behavior. The dynamic behavior is introduced into the diagram by appending an arrow with the letter **t** (time).

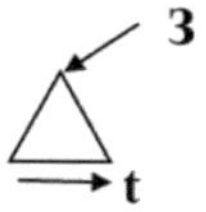

$$\text{(8.9)}$$

We note that B. Zlotin[1] introduced process analysis (a kind of dynamic Su-Field analysis). B. Shmakov[2] suggested having time-dependent Su-field analysis.

8.1.1   Patterns of the EAK Development

The EAK development is carried out according to a law similar to the Su-field increasing degree law (Fig. 8.7).

EAK development starts when the system does not incorporate any knowledge (K), i.e. regular El-Action development (Figs. 7.1, 7.2, 7.3, and 7.4). We name this stage as **non-EAK system**.

A **simple EAK** is the stage in which Knowledge (K) is introduced into the system. The next stage is the increased controllability of the system, i.e., we can **change and coordinate the Elements (E), Action (A), and Knowledge (K)**.

---

[1] Zlotin B. Process Analysis.—Leningrad, 1977 (Russian).

[2] Shmakov B. V. and other. Su-Field Analysis of the Technical Systems. Textbook for the course "Theory of solving engineering problems"/B. V. Shmakov, P. D. Krikun, E. G. Schepetov, ed. F. Y. Izakov.—Chelyabinsk: ChPI, 1985. 58 p. (Russian).

Fig. 8.7 The general trend of EAK

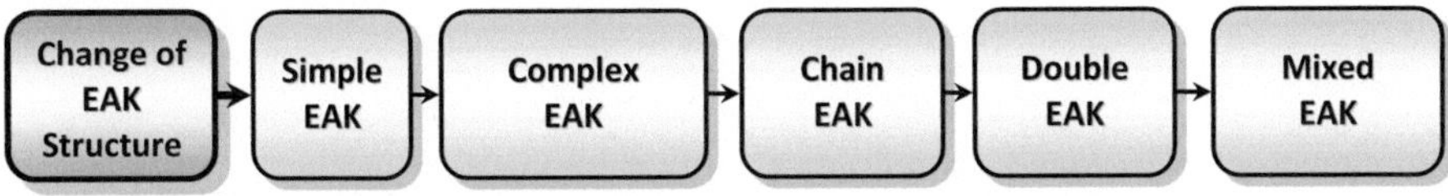

Fig. 8.8 The trend changes of EAK structure

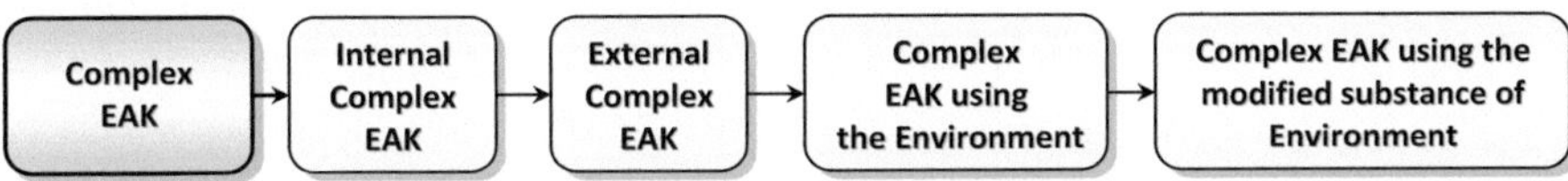

Fig. 8.9 Complex EAK change trend

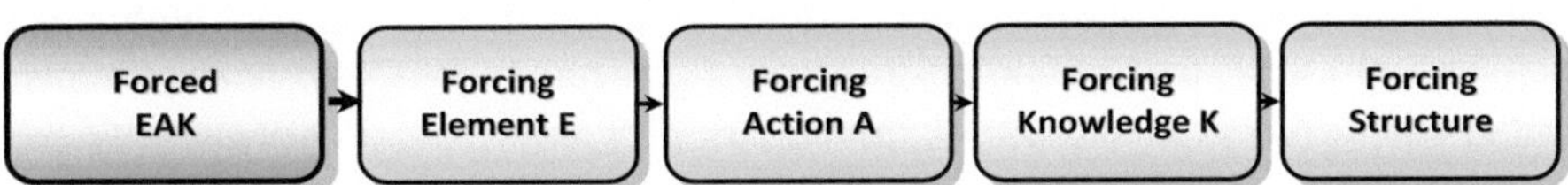

Fig. 8.10 Forced EAK change trend

Coordination means that Action (A) responds to the Knowledge (K) and Element (E) responds to the Action (A). Hence, the Knowledge (K) identifies the changes in the Elements (E) and Actions (A) by controlling them.

The EAK structure is a modified form of the structures shown in Figs. 8.8, 8.9, and 8.10 obtained by the introduction of knowledge (K).

8.2 Parametric Analysis

It is necessary to take into account all the parameters of the constituent components (element, action, and knowledge) to complete the picture.

8.2.1   Element Data

Element Data consists of

1. Structure
2. Properties
3. Changes in time

Element Structure

Element structure is understood as

1. Internal structure and/or composition
2. Shape
3. The aggregate state. Use the trends of the Degree of Fragmental Increase (Fig. 6.7).

   The structure depends on the element itself.

Example 25. Mechanical system structure

In any system design, system elements must be described, the element form (design), composition material (plastic, metal, "smart" substance, e.g., material with shape memory), physical state, its composition (plastic, metal, "smart" substance, Element shape) and its physical state (solid, liquid, gas, plasma).

Example 26. Organizational system structure

For organizational systems, the company or division structure must be explained.

Example 27. Business system structure

A business system, besides needing the organizational structure, business goals, and means should be described.

Example 28. Information system structure

For information systems (not company, or business), this may include the structure of the element, the type of information and its parameters, etc.

Element Properties

Elements description includes all of the properties and parameters of the element. It can be:

– Technical properties, including weight and dimensions
– Economic characteristics
– Aesthetic characteristics
– Ergonomic characteristics
– Environmental performance
– Psychological characteristics

Time Changes

Time changes take into account whether the element and its characteristics change with time, that is, the element is dynamic or static and the characteristics of dynamics.

8.2.2 Action Data

As data can be considered as an element we are interested in the following:

1. Kind of action
2. Gradient action (grad A).

Kind of Action

Any action, impact, and interaction of any *kind*:

1. Flows
2. Substances
3. Fields
4. Information
5. Forces
6. Energy

Action Gradient

Gradient (from the Latin. Gradiens, genus. Case gradients—walking, growing) is a vector, with its direction indicating the direction of the greatest increase of some scalar quantity ψ, the value of which varies from one point of space to another (scalar field) and the value (modulus) equal to the growth rate of this value in this direction.[3]

$$\mathbf{grad}\,\psi = \frac{\partial^2 \psi}{\partial x^2} i + \frac{\partial^2 \psi}{\partial y^2} j + \frac{\partial^2 \psi}{\partial z^2} z \tag{8.10}$$

The gradient is a vector, with its direction indicating the direction of the fastest increase of some quantity ψ.[4]

Therefore, for action A we get:

Considering that the quantity we are dealing with is action (A), formula (8.10) can be represented as

$$\mathbf{grad}\,A = \frac{\partial^2 A}{\partial x^2} i + \frac{\partial^2 A}{\partial y^2} j + \frac{\partial^2 A}{\partial z^2} k \tag{8.11}$$

---

[3] Gradient—Wikipedia (in Russian).

[4] Gradient function URL: https://math.semestr.ru/math/gradient.php (in Russian).

Obviously, for action A, we define i, j, k, as:

- *i—direction*
- *j—strength*
- *k—speed*

In some cases, we need to specify higher derivatives and an integral of the *action*.

8.2.3   Knowledge Data

Some definitions.[5]

Knowledge in the widest sense is a set of concepts, theoretical constructions, and ideas.

Knowledge is a subjective image of objective reality, that is, a reflection of the external and internal world in the mind of the beholder in the form of presentations, concepts, judgments, and theories.

Knowledge includes what exists and is observed. Knowledge is crucial for the ordering and systemizing results of human activity. There are various types of knowledge: scientific, ordinary (common sense), intuitive, religious, etc. Ordinary knowledge is the basis for the orientation of a person in the outside world, the basis of his daily behavior and foresight, but usually it contains errors and contradictions. Scientific knowledge is logical, reproducible verifiable, and evidence based. This approach to science helps eliminate errors and overcome contradictions.

The **distinctive characteristics of knowledge** are still subject to uncertainty in philosophy. According to most thinkers for something to be considered knowledge is something that must meet three criteria:

- be confirmed
- true
- trustworthy.

Knowledge—facts, information, and skills acquired through experience or education; the theoretical or practical understanding of a subject.[6]

As knowledge data can be considered:

1. Structure
2. Properties
3. Dynamic that is it may vary with time

---

[5] **Knowledge (in computer science)**. URL: https://dic.academic.ru/dic.nsf/ruwiki/929992 (in Russian).

[6] https://www.lexico.com/en/definition/knowledge.

Knowledge Structure

Knowledge comprises of

1. Type of knowledge
2. Components of knowledge and their interactions

Knowledge Properties

By knowledge properties, we mean what can be extracted from this knowledge.

Dynamic that is It May Vary with Time

Knowledge may be

1. Unchanged
2. Changing, for example, adaptable

8.3 Patterns of Control Element

The pattern control element is an integral part of *the law of control and dynamics changes degree*. It belongs to the group of the *evolution systems laws* (Fig. 8.11).

A Control Element uses a "**smart**" element, changing **the concentration** and **connectedness of that element**.

Connectedness in physical systems includes all types of connections. That is, the connection between the parts of the object, the particles of matter, and up to the inter-molecular and atomic bonds.

In such physical systems, **Connectedness** can increase or decrease the *stiffness* or *elasticity* of the object along with the *number of degrees of freedom* in the system. Whereas in IT or software systems, connectedness refers to the relationship between parts of the element.

The pattern of control **element** is *the adaptation of internal and external changes*, i.e. the *changes in <u>controllability</u> and <u>dynamics</u>*.

The pattern of the control element carried out in (Fig. 8.11) contains the following trends:

– *Use of "smart" elements*
– *Changes in the degrees of the freedom number*
– *Changes of an element concentration*
– *Changes in the fragmentation degree*
– *Transition to capillary porous materials (CPM)*
– *Increase in the emptiness degree*

Increasing the emptiness degree is a special case of the *transition to the CPM*, and the transition to the CPM is a special case of *changing the degree of fragmentation*.

Connectedness includes the following: *changing the number of degrees of freedom* number, *changing the degree of fragmentation*, the transition to *the CPM*, and *the increase in the emptiness* degree (Fig. 8.1).

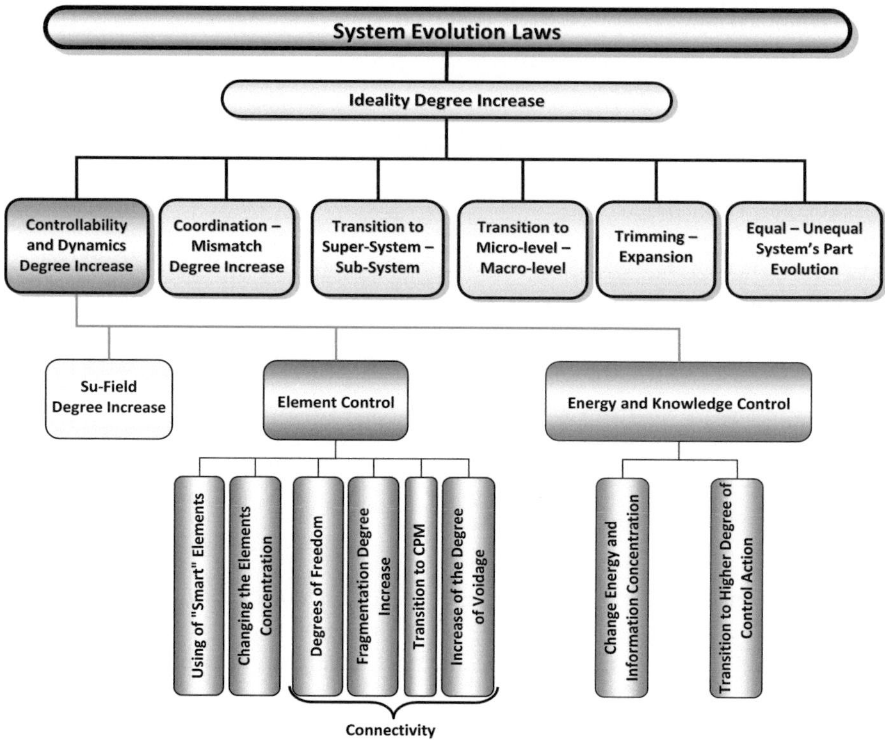

Fig. 8.11 The structure of the technical systems evolution law

8.4   Energy and Information Control

The pattern of energy and information control is an integral part of the *law of controllability and dynamics increase*. The latter refers to a group of the *systems' evolution laws* (Fig. 8.11). The pattern is that *any system in its development tends to change the energy and information density at the proper time and place*.

Let us consider the mechanisms of energy and information density, which are primarily related to a working unit.

Energy and Knowledge Control is carried out by the trends (Fig. 8.12):

- **Change Concentration**

  - Energy
  - Information

- **Transition to Higher Degree of Control Action**

  - Replacing the Action Types
  - Transition to MONO-BI-POLY Action
  - Dynamic Action

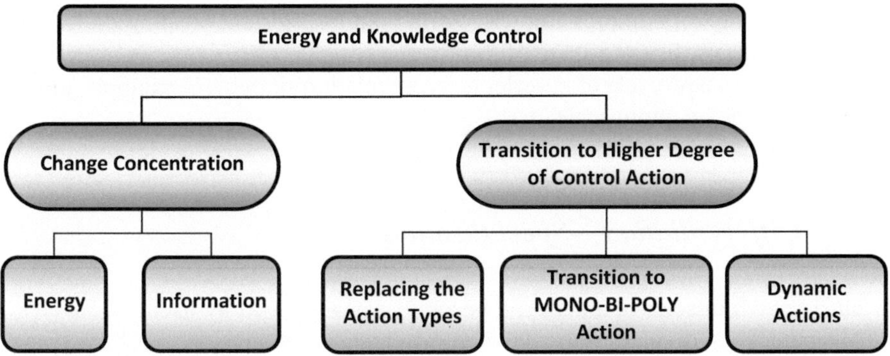

Fig. 8.12 Energy and knowledge control law

8.5 Patterns of Knowledge Development

We have identified the following patterns of knowledge development:

– The expansion–compression (convolution)
– Differentiation–specialization
– Combination of known knowledge and its integration
– Intellectualization.

8.5.1 Extension—Compression (Convolution)

The tendency of expansion–compression is illustrated by the development of different theories.

Example 29. The development of the electromagnetism theory

First, electricity and magnetism were considered as two separate forces. Then, many scientists have noticed the connection of electrical and magnetic phenomena. This is the stage of ***knowledge expansion***.

James Maxwell has linked together the electrical and magnetic phenomena, creating classical electrodynamics. This is the stage of ***knowledge compression*** (*the four Maxwell Equations*).

Example 30. The development of the gravitational theory

Copernicus, Kepler, and Galileo represent the stage of ***knowledge expansion***.

Isaac Newton discovered the *law of universal gravitation*. It was a stage of ***knowledge compression.***

Further accumulation of knowledge (***extension***) showed the inaccuracy of the Newtonian theory.

The next stage of compression was carried out by Einstein creating *the general theory of relativity*. It was a stage of **knowledge compression.**

Then again, a new cycle of knowledge accumulation started, for example, gravity processes on the quantum scale.

8.5.2   Differentiation—Specialization

When a branch of science matures, it becomes a new science by itself.

Mechanics is divided into classical mechanics, relativistic mechanics, and continuum mechanics. The latter is divided into the fluid mechanics, acoustics, and mechanics of solids.

Each of the sections continues to divide and specialize further.

8.5.3   Combining of the Known Knowledge and Its Integration

Combining the knowledge from a few separate fields and generating a unified theory that can better explain certain phenomena.

For example, physics and chemistry are well-known sciences, and the combined physical chemistry studies macro chemical behavior with thermodynamics.

Altshuller[7] described a few methods of knowledge combination.

8.5.4   Intellectualization

The transition from uncontrolled to controlled knowledge is due to the following chain: *adaptive (self-adjusting)* knowledge, *self-learning* and *self-organizing* knowledge, *self-evolving*, and *self-reproducing* knowledge.

To date, there are adaptable, self-adjusting, and self-learning systems, able to adapt and accumulate knowledge in learning. The development of artificial intelligence should lead to self-evolving and self-perpetuating knowledge.

This is a pattern of the future of knowledge development.

---

[7] Altshuller G. S. How to make discoveries. Thoughts on the methodology of scientific work.—Baku, 1960 (Russian) http://www.altshuller.ru/triz/investigations1.asp.

8.5.5 *Example*

Consider the process of making chocolate (Fig. 8.13).

At first, the process was carried out manually. The person knew the whole process: choosing the right cocoa beans, roasting, and grounding them to the desired consistency. The knowledge about the process of making chocolate was only in the employee's head, that is, knowledge was not present in the system.

$$A \to E \tag{8.12}$$

where

A Action
E Element

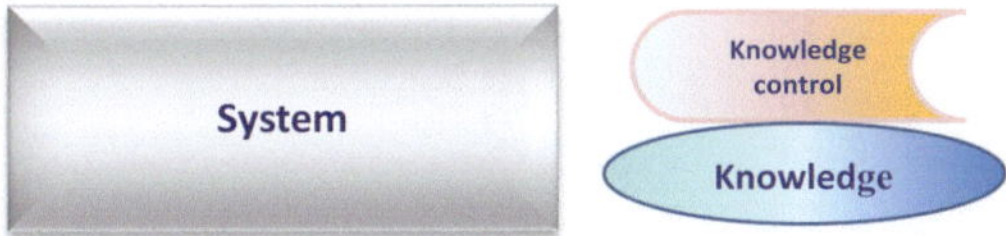

The next step introduced simple mechanisms and machines. The machines had some knowledge, for example, a machine (mill) to crush the cocoa beans. By this step, some knowledge was incorporated into the system (Fig. 8.14).

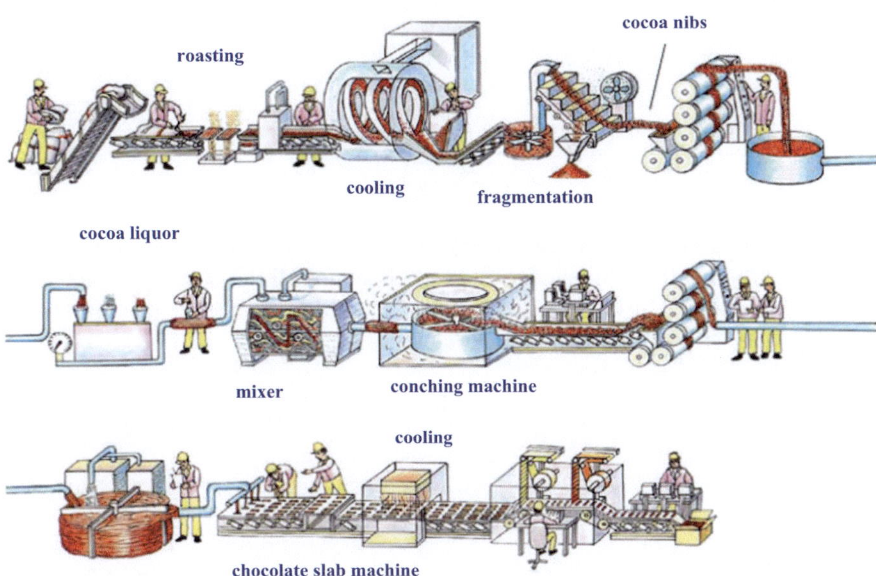

Fig. 8.13 The process of making chocolate

Fig. 8.14 A machine (mill) to crush the cocoa beans

$$\begin{array}{c} \mathbf{K_1} \\ \diagdown \\ \mathbf{A} \longrightarrow \mathbf{E} \end{array}$$

(8.13)

where

K₁   Knowledge
A   Action
E   Element

Then automation was introduced and even more knowledge was added to the system. Finally, a fully automated system that contains all the necessary knowledge for chocolate manufacturing was created. Control of this knowledge (programming) is outside the system in the programmer's mind (Fig. 8.15).

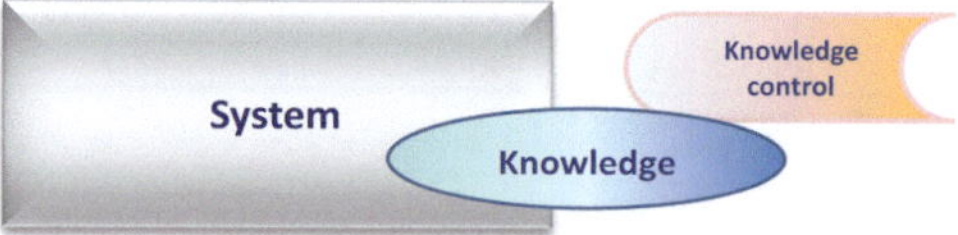

$$\begin{array}{c} \mathbf{K_1} \\ \diagdown \\ \mathbf{A} \longrightarrow \mathbf{E} \end{array}$$

(8.14)

Fig. 8.15 A fully automated system

where

K₁ Knowledge
A   Action
E   Element

The next step in the development requires the introduction of knowledge control of chocolate manufacturing.

For example, the system will adapt and modify the process for different grades of cocoa and, particularly, adjust itself to the different cocoa beans present in the system. The system will learn the process of chocolate making and improve it. The system will reconstruct itself for different types of chocolate. The system will self-develop and create new chocolate recipes. The system will create a similar system itself.

$$(8.15)$$

where

$\mathbf{K_1}$ knowledge, control action
$\mathbf{K_2}$ the knowledge of the Element (E) state
$\mathbf{K_3}$ knowledge that controls knowledge

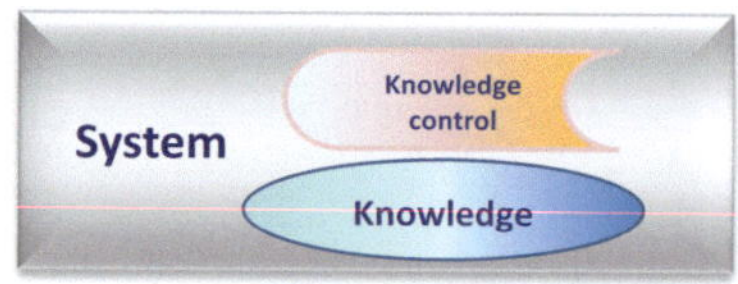

8.6   Structural Analysis of Information Processing Systems

8.6.1   Definition

Consider the features of the application for EAK analysis of information processing systems.

In information processing systems, we deal with **Data** and **Functions**. In these systems, an *Element* is represented as **Data (D)** and *Action*—a **Function (F)**.

A model that includes the **Data**, **Function**, and **Knowledge** is called **DFK**. Method of analysis and transformation DFK will be called **DFK analysis**.

– **Data**—incoming information to be processed by the system
– **Function**—processing applied to the incoming information
– **Knowledge**—aggregation of all proven, empirical, or other a priori information related to how the data is to be processed.

The main difference between Knowledge and Data is in structure and availability. The knowledge is defined during the design or update of the system and exists independently of the incoming Data. When new facts or relations are discovered, the Knowledge base is updated and the processing is accommodated to include the new understanding. In certain cases, past Data forms part of the Knowledge available for processing the new Data.

Data, Function, and Knowledge are parts of DFK, and the corresponding methodology is called DFK Analysis.

8.6.2 *DFK Analysis*

The system is called uncontrolled if its function is fixed and does not depend on knowledge. Such a system is referred to as an **Incomplete DFK** (2.3):

$$F \to D \tag{8.16}$$

where

F Function
D Data

Usually, some *prior **knowledge*** about the incoming data is available and used for adjusting the functionality of the system accordingly. Such a system forms a **Simple DFK**, shown in (8.17):

$$\tag{8.17}$$

where

K Knowledge
F Function
D Data

The system may adapt its functionality autonomously, by analyzing the incoming data and selecting the best approach to process it. This is a **Complete DFK** (or **DFK**), presented in (8.18):

$$\tag{8.18}$$

where

K Knowledge
F Function
D Data

We believe that the proposed concept is helpful for the analysis of the effectiveness and improvement of the existing solution as well as for the design of the new ones.

Example 31. Data compression system

Let's consider a system for data compression.

1. When the input data type is unknown, the only reliable approach is lossless compression methods that achieve relatively low compression ratios. This system employs no knowledge, and therefore it is an incomplete DFK (8.19).

$$\mathbf{F} \rightarrow \mathbf{D} \tag{8.19}$$

where

F Function
D Data

2. When the data type is known, e.g., image or audio, specific compression schemes can be applied for that type of data, like JPEG for all images or MP3 for all audio streams. Dedicated compression schemes make use of data structure to achieve much higher compression performance. The system effectiveness is higher compared with the first example. This system uses only external knowledge provided from outside, without any analysis of the input, and there it is a simple DFK (8.20).

$$\tag{8.20}$$

where

K Knowledge
F Function
D Data

3. To achieve optimal compression performance for a particular input, an image compression system analyzes the input and determines the type of image (e.g., photo, drawing, text, and medical) to select an optimal method for this specific type of image. This system uses both the external knowledge provided from outside and internal knowledge collected by analysis of the input data and therefore it is an ***adaptive DFK*** (8.21).

$$\tag{8.21}$$

where

K Knowledge
F Function
D Data

8.6.3 Laws of DKF Degree Increase

Information processing systems tend to increase their degree of DKF according to the laws. We state here three laws of DKF degree increase:

1. **Law of multistage processing**
2. **Law of multiple source processing**
3. **Law of accommodation**

The law of multistage processing

The law of multistage processing states: any information processing system tends to process data in stages. That is, as the processing complexity rises, processing is divided into several stages. There are a number of distinct reasons for multistage processing:

1. *Distributed systems.* The systems where the information processing is performed by different components.

Example 32. World wide web

On the World Wide Web, the information is distributed among many servers. Processing of the information is done on both the servers and client computers.

2. *Development optimization.* Complex systems are divided into components, so each component can be developed and verified independently. Minimizing the interfaces between the components makes system development more efficient and faster.

Example 33. Component separation

All processing systems are divided into components, e.g., computer programs are divided into functions.

According to the law of multistage processing, **Simple DFK** becomes **Simple Multistage DFK**. In Simple Multistage DFK, each processing stage is independent from all other stages, only data is shared. The **Simple Multistage DFK** transforms into **Coordinated Multistage DFK**, where a partial amount of knowledge is being shared between the stages. Finally, **Coordinated Multistage DFK** becomes **Common Multistage DFK**. In Common Multistage DFK, all knowledge is fully shared between all processing stages (Fig. 8.16).

1. *Simple Multistage DFK*

Example 34. Webpage image

The image shown on a web page taken by the camera is presented on a monitor. The camera and monitor do not share any information. The photo shown on a web page was generated by a camera. Camera is capturing light information, processes it, and

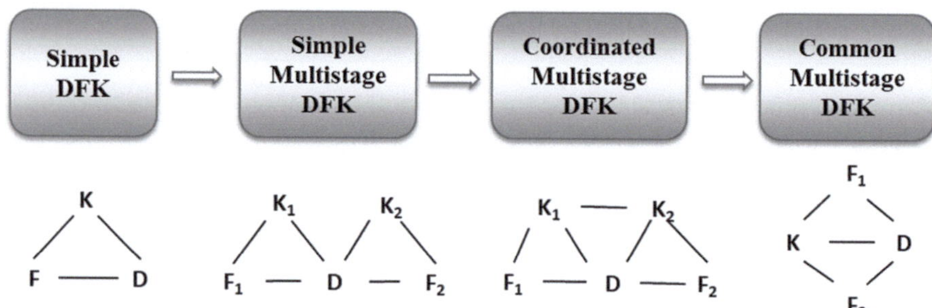

Fig. 8.16   Trend of multistage processing

Fig. 8.17   The image shown on a web page

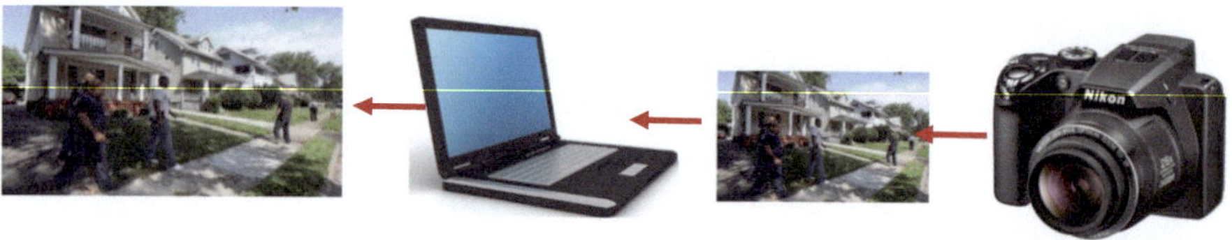

Fig. 8.18   The process produces the final photos

creates an image file. Computer that presents the web page may further process the image file, for example, to adapt its color and size for the best presentation on the monitor. Camera and computer may not share the information (Fig. 8.17).

2.   *Coordinated Multistage DFK*

There are many cases, when along with the information metadata is included.

Example 35. DSLR Camera

DSLR camera captures raw (partially processed) images along with the capture settings. After transferring the raw images to the PC, these settings are used to complete the processing and produce final photos (Fig. 8.18).

3.   *Common Multistage DFK*

Processing algorithms divided into subroutines can have a shared structure to contain all the knowledge. During the processing of each input, the global knowledge of the system is used (Fig. 8.19).

The law of multiple sources processing

The law of multiple sources processing states: an information processing system with multiple sources tends *to process the multiple sources jointly*. That is, multiple inputs with the same or different sorts of information can be jointly processed to explore the correlation between them.

Example 36. Multiple sources of information

Multiple sources of information are video capture, where both visual and audio data are captured. The relation between them can be exploited to improve speech recognition, noise reduction, and video compression (Fig. 8.20).

Fig. 8.19   The process of input to the global knowledge

Fig. 8.20   Video camera

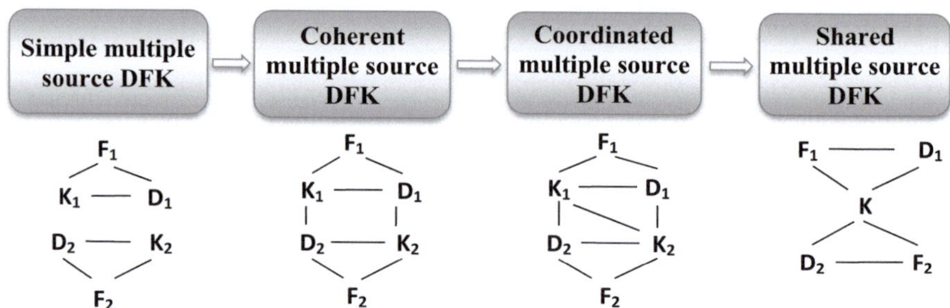

Fig. 8.21   Trend of multiple sources processing

Fig. 8.22   Video conferencing

According to the law of multiple sources processing, any **Simple Multiple Source DFK** tends to become **Coherent Multiple Source DFK**. In **Coherent Multiple Source DFK** all sources of information are observed and used for processing, but the processing is performed independently. **Coherent Multiple Sources DFK** may transform to **Coordinated Multiple Sources DFK**, where knowledge is partially shared by the processing subsystem on each source. Finally, **Coordinated Multiple Sources DFK** may fold into the **Shared Multiple Sources DFK** with centralized knowledge handling (Fig. 8.21).

Example 37. Videoconferencing

1. *Simple multisource DFK*

Video conferencing with two independent systems: audio and video transmission (Fig. 8.22).

2. *Coherent multisource DFK*

In video conferencing, the directional microphone is focused on the speaking person based on the processing of the video signal (Fig. 8.23).

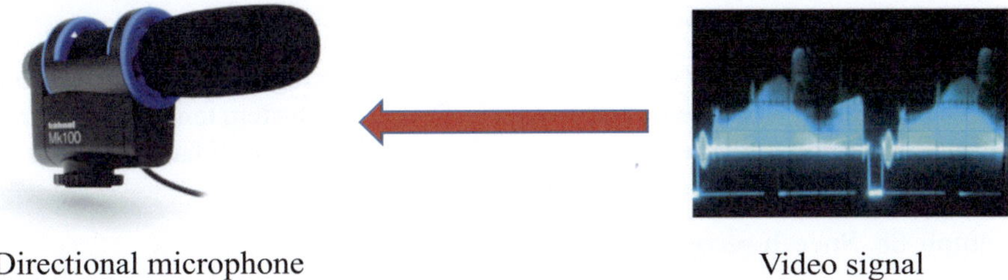

Fig. 8.23 Directional microphone is focused on the speaking person

3. *Coordinated multisource DFK*

In videoconferencing, video is zoomed using the video processing and direction of arrival of the voice, while the microphone is focused based on the audio processing and location of the speaker person. This may be especially useful to identify the main speaker when there are several people talking in the background (Fig. 8.24).

4. *Shared multisource DFK*

In videoconferencing, video is zoomed and the directional microphone is focused based on the joint analysis of video and audio channels (Fig. 8.25).

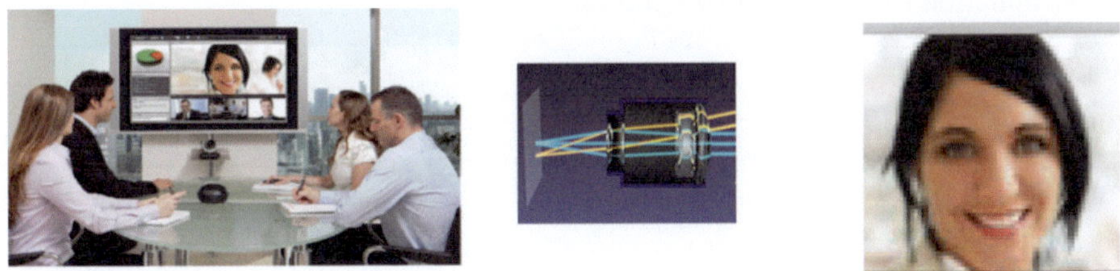

Fig. 8.24 Location of speaker person

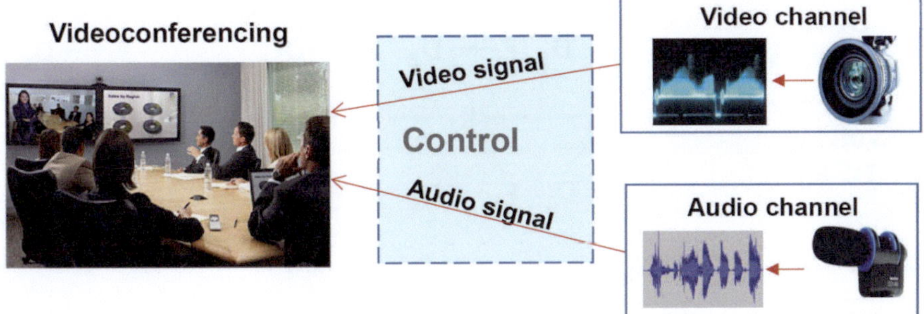

Fig. 8.25 The joint analysis

The law of accommodation

The law of accommodation states: an information processing system tends to accommodate past data to improve its performance. That is, the system tends to learn from information received in the past and adjust itself to produce optimal results for the incoming information.

Example 38. Speech recognition system

Unsupervised learning systems, like speech recognitions and search engines.

   Static DFK is not changing in time, having constant a priori set knowledge (K) and functionality (F). Only the input data (D_1–D_n) is changing over time. Static DFK may become **Learning DFK** if the knowledge base (K_1–K_n) changes with the incoming data (D_1–D_n). When not only the knowledge (K_1–K_n) but also the function (F_1–F_n) applied to the data (D_1–D_n) changes with time such a system is an **Evolving DFK** (Fig. 8.26).

Example 39. Speech recognition system

1. *Static DFK.*

    - Only a fixed predefined dictionary is used for speech recognition.
    - Different data ($\mathbf{D_1}$–$\mathbf{D_n}$) is presented every time, but the knowledge ($\mathbf{K}$) and function ($\mathbf{F}$) remain the same.
    - Knowledge ($\mathbf{K}$) is a fixed predefined dictionary.
    - Function ($\mathbf{F}$) is speech recognition.

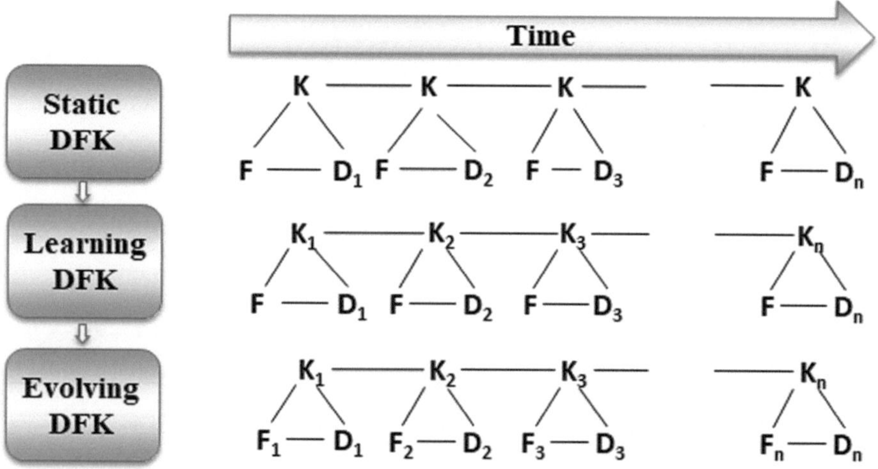

Fig. 8.26 Trend of accommodation

K – const.
Fixed dictionary

F – const.

D – var.

2. *Learning DFK.*

- The system is implemented as unsupervised learning systems.
- Every time the speech is recognized with high confidence, the dictionary is retrained to include the new sample.
- Over time, the dictionary becomes tailored to the particular speakers that use the system.
- System knowledge $(\mathbf{K_1}–\mathbf{K_n})$ adapts to the environment but the function $(\mathbf{F})$ is not changed.

K – var.
Adaptive dictionary

F – const.

D – var.

3. *Evolving DFK.*

- In evolving DFK the system not only adapts its knowledge $(\mathbf{K_1}–\mathbf{K_n})$ but also modifies its function or adds new capabilities.
- The speech recognition system may add functionality $(\mathbf{F_1}–\mathbf{F_n})$ of talking back and synthesizing the speech based on the collected knowledge.

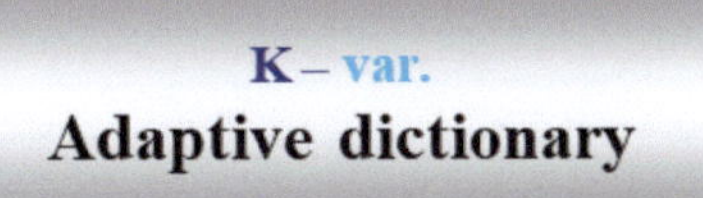

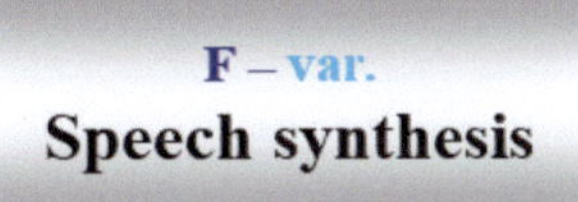

D – var.

Complex DFK

The laws presented above are also applicable in combination. The example below shows Common Multistage Shared Multiple Source Learning DFK.

There are two sources of information D_{1n} and D_{2n}. Shared Knowledge K_n learns data behavior with each time sample. Function F_{1n} applied on D_{1n} is evolving, so the $D_{1n}–F_{1n}–K_n$ is an Evolving DFK. On the source D_{2n} multistage processing is applied by the functions F_2 and F'_2 with a common learning knowledge K_n (Fig. 8.27).

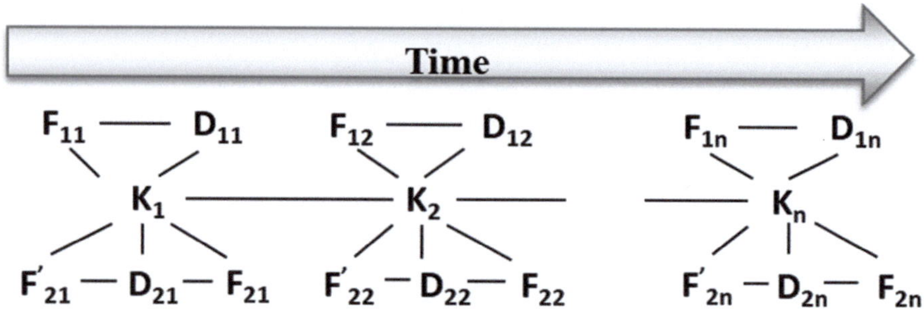

Fig. 8.27 Example of complex DFK

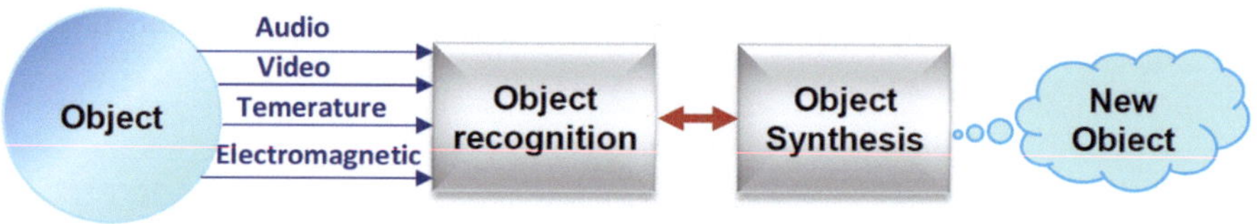

Fig. 8.28 The system synthesizes new object

Example 40. Object recognition

The system may recognize objects not only by audio or visual information, but also by smell, temperature, electromagnetic radiation, etc.

In addition, the system may synthesize objects with given parameters (Fig. 8.28).

Chapter 9
Independent Work

9.1 Quiz

1. What is Su-field analysis? Give a definition.
2. What is Su-field? Give a definition.
3. What is the substance in Su-field analysis? Give a definition. What are the kinds of substances?
4. What is the field in Su-field analysis? Give a definition. What are the kinds of fields?
5. What is the purpose of Su-field analysis? What are the uses of Su-field analysis? Presentation of original problem structure. Getting a structural solution to the problem. Prospects of development of system structure.
6. What is the kind of interaction (basic configuration)?
7. What are the types of Su-field systems?
8. What are the types of Su-field models for fields?
9. What are the kinds of Su-field structures. Call them.
10. What is a generator of fields? Give an example.
11. What is a converter of fields? Name types of field converters. Give an example.
12. What are the features of "measuring" Su-field model?
13. What is a non-Su-field system? Give a definition.
14. What are the types of Su-field models? Name them.
15. What is an Incomplete Su-field system? Give a definition. Give an example.
16. What is a Complex Su-field? What are the types of Complex Su-field? Give an example.
17. What is an Internal Complex Su-field? Give an example.
18. What is an External Complex Su-field? Give an example.
19. What is a Complex Su-field in an external environment? Give an example.
20. What is a Complex Su-field in a modified external environment? Give an example.
21. What is a Compound Su-field? What are the types of Compound Su-field?

V. Petrov, *Structural System Analysis*, https://doi.org/10.1007/978-3-031-55825-2_9

22. What is a Chain Su-field?
23. What is a Double Su-field?
24. What is a Mixed Su-field?
25. What are the types of elimination of harmful interactions? Give an example.
26. Find a technological effect.
27. What is the law of increasing degree of Su-field? The main trend in the development of Su-fields. The trend of change in the Su-field model.
28. What are the specifics of using Su-field analysis for information systems?
29. What is the difference between a new representation of a Su-field analysis?
30. What is the name of Su-field analysis for information systems?
31. What are the patterns of knowledge development?
32. Describe the patterns of development of a new type of Su-field analysis for information systems.
33. Describe the patterns of development of information processing systems.

9.2 Topics for Reports and Abstracts

1. History of the development of Su-field analysis. Show changes in Su-field analysis.
2. Trends of Su-field analysis.
3. Su-field future analysis.
4. Su-field analysis on information technology.

9.3 Assignment

1. *Give an example of different types of substance (element), field (action), and knowledge*

    1.1. Give an example of different substances (elements).
    1.2. Give an example of different fields (action).
    1.3. Give an example of different knowledge.

2. *Build a Su-field model for examples. Present condition and solution in the form of Su-field*

    2.1. Examples 12, 14.
    2.2. Example 41. Turbine of the jet engine.

The turbines of jet engines operate at high-temperature values. In order to preserve the strength characteristics of turbine blades, it is necessary to add alloy additives to the source material, for example, cobalt, which significantly increases the cost of the turbine, however, confers resistance to high temperatures to it. Pratt and Whitney developed a technology for blade manufacturing, which enables to reduce the content

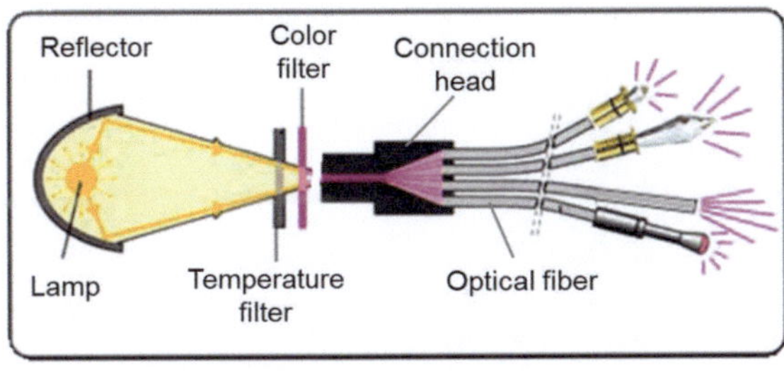

a) Structure of lighting fixture

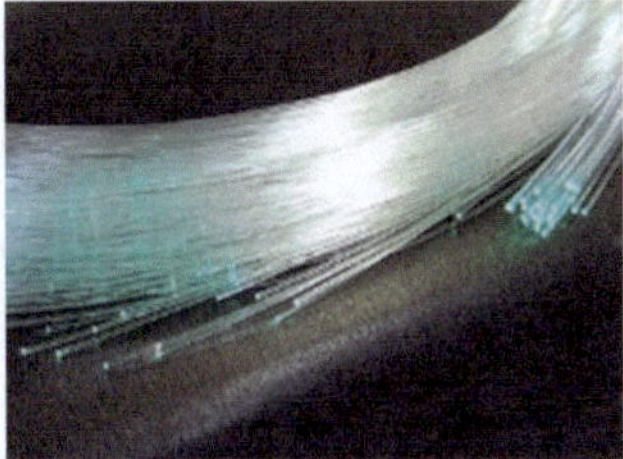

b) Light from beam mode waveguides

Fig. 9.1 Decorative lighting fixture made of optical fiber

of cobalt in them by 30%. To achieve this, they drill very small holes in the blades with a laser. The air passing through the holes, cools the blades, and besides, aerodynamic resistance is reduced. Thus, the turbines can be manufactured of a material, which is less heat-resistant.[1]

2.3.   Example 42. Elimination of cavitation

Cavitation causes erosion (destruction) of the material of the devices, where it occurs. Attempts are made to eliminate cavitation. In this case, it is very important that the cavitation should be eliminated uniformly. It is suggested to act upon cavitation bubbles with ultrasound oscillations in the range of frequencies from 1 to 50 kHz.[2]

2.4.   Example 43. Power measuring

Calorimetric method for measuring power. In order to measure power absorbed by the load in the super-high-frequency (SHF) range, the amount of heat is determined, which is given by the load to the operating fluid (water), and the working medium is used as a load. With the aid of the measuring unit, the temperature of the working medium is registered and its value is used for determining the value of power.[3]

2.5.   Example 44. Decorative lighting fixture

Decorative lighting fixtures are known, which use optical fiber. Such a lighting fixture (Fig. 9.1) consists of a lamp, reflector, temperature filter, and color filter, connecting head, and optical fiber cable. In this lighting fixture, there was one color filter, which was rigidly fastened.

Compile a Su-Field formula.

A decorative lighting fixture was invented, which changes its color with the variation of atmospheric pressure (Fig. 9.2). In this invention, the color filters are fastened on

---

[1] Inventor and rationalizer, No. 12, 1985, Micro Information 1203, p. 30.

[2] Author's certificate 954,597.

[3] Yelizarov A. S. Electric radio measurements. – Minsk: Vysheishaya shkola, 1986, 320 p.

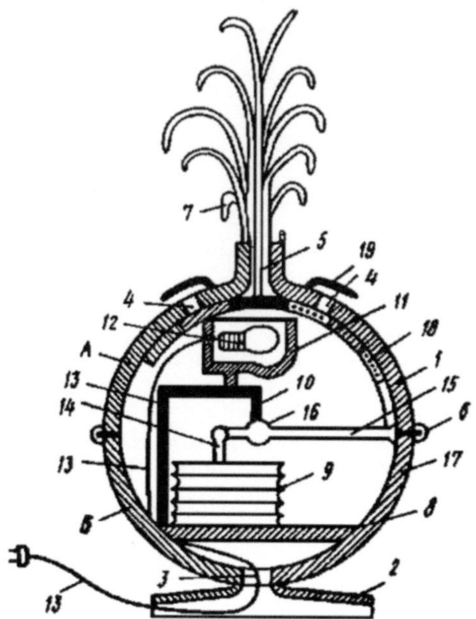

Fig. 9.2 Decorative lighting fixture. Author's certificate 779,726. 1—spherical body made of heat resistant plastic, consisting of upper hemisphere A and lower hemisphere B; 2—support; 3—orifice in the lower part of the body; 4—ventilation orifice; 5—throat; 6—annular bracket, connecting halves A and B; 7—beam mode waveguides, built into the throat 5; 8—the substrate; 9—hermetically sealed corrugated vacuum chamber; 10—holder; 11—reflector; 12—source of light; 13—electric power cable; 14—guide path; 15—lever; 16—hinge; 17—fastening of the lever; 18—color filter (light conducting plate, subdivided into individual color sectors); 19—light-reflecting protective screens

Fig. 9.3 Safety line

a corrugated vacuum chamber, which changes its volume depending on atmospheric pressure and moves color filters of different colors.[4]

2.6.   Example 45. Safety line

A safety line of alternating rigidity is proposed.[5] The line has a loop, which is connected by a ligament, which has a lesser tensile strength than the line (Fig. 9.3). If a person falls down, the ligament is ruptured first, extinguishing part of the energy of falling.

---

[4] Author's certificate 779,726.

[5] Author's certificate 631 631. Loop of balance. Inventor and rationalizer, 11/79, P. 28.

Fig. 9.4 Line with a shock absorber

2.7. Example 46. Line with a shock-absorber

When the wind is strong the line connecting anchor with a boat is ruptured. M. Lobov from Pensa proposed (similarly to the previous example) to make a loop out of a line and to connect it with a rubber ribbon (Fig. 9.4). Such loops are made in the place of fastening the line to anchor and boat.[6]

2.8. Example 47. Power transmission line

Power transmission line with a section of cable, forming a loop, in parallel to which an element is installed, which prevents the wire from rupture under overrated mechanic loads, in which the element, preventing the wire from rupture (for the purpose of simplifying the line structure) is embodied as a bridge with mechanical strength, which is lower that the mechanical strength of the line cable.[7]

3. *Determine what kind of knowledge it is*

3.1. Example 48. The inflow and outflow of cold molecular gas

Observations by ALMA and data from the MUSE spectrograph on ESO's VLT have revealed a colossal fountain of molecular gas powered by a black hole in the brightest galaxy of the Abell 2597 cluster—the full galactic cycle of inflow and

---

[6] Young technician, 1/73, P. 42.

[7] Author's certificate 433,580.

outflow powering this vast cosmic fountain has never before been observed in one system.

A mere one billion light-years away in the nearby galaxy cluster known as Abell 2597, there lies a gargantuan galactic fountain. A massive black hole at the heart of a distant galaxy has been observed pumping a vast spout of cold molecular gas into space, which then rains back onto the black hole as an intergalactic deluge. The in- and outflow of such a vast cosmic fountain has never before been observed in combination, and has its origin in the innermost 100,000 light-years of the brightest galaxy in the Abell 2597 cluster.

"This is possibly the first system in which we find clear evidence for both cold molecular gas inflow toward the black hole and outflow or uplift from the jets that the black hole launches," explained Grant Tremblay of the Harvard-Smithsonian Center for Astrophysics and former ESO Fellow, who led this study. "The supermassive black hole at the center of this giant galaxy acts like a mechanical pump in a fountain."

Tremblay and his team used ALMA to track the position and motion of molecules of carbon monoxide within the nebula. These cold molecules, with temperatures as low as minus 250–260 °C, were found to be falling inwards to the black hole. The team also used data from the MUSE instrument on ESO's Very Large Telescope to track warmer gas—which is being launched out of the black hole in the form of jets.

"The unique aspect here is a very detailed coupled analysis of the source using data from ALMA and MUSE," Tremblay explained. "The two facilities make for an incredibly powerful combination."

Together these two sets of data form a complete picture of the process; cold gas falls towards the black hole, igniting the black hole and causing it to launch fast-moving jets of incandescent plasma into the void. These jets then spout from the black hole in a spectacular galactic fountain. With no hope of escaping the galaxy's gravitational clutches, the plasma cools off, slows down, and eventually rains back down on the black hole, where the cycle begins anew.

This unprecedented observation could shed light on the life cycle of galaxies. The team speculates that this process may be not only common, but also essential to understanding galaxy formation. While the inflow and outflow of cold molecular gas have both previously been detected, this is the first time both have been detected within one system, and hence the first evidence that the two make up part of the same vast process.[8]

3.2. Example 49. Genetics

In accordance with the provisions of classical genetics, the properties of living organisms acquired during life are not inherited unless the structure of the DNA located in the nucleus of the corresponding cells responsible for heredity is changed. However, this situation could not explain the successes of the practical hybridization of animals and plants, allowing them to obtain new and heritable properties of organisms in a

---

[8] ALMA and MUSE Detect Galactic Fountain. Science Release. 6 November 2018 https://www.eso.org/public/news/eso1836/.

relatively short time. The DNA structure in the cell nucleus is very stable and poorly mutated.

In recent years, advances in molecular and cellular biology have accounted for this paradox. It turns out that in the cells of the body outside the nucleus (in the cytoplasm) there is also DNA, as well as informational RNA, which is more sensitive to external influences and at the same time, can affect heredity. One of the most unusual manifestations of RNA interference and RNA silencing is that they make possible an unheard-of thing from the point of view of classical genetics—the inheritance of acquired characters. Interference and silencing do not alter the sequence of genes in DNA, but can control how certain genes are active. Indeed, it is easy to imagine that if regulatory RNA gets into progeny cells from an egg cell, they will be able to bring with them a certain pattern, pattern of gene activity. Moreover, as it turns out, this pattern is able to be inherited for several generations.

4. *Define the kind of DFK*

4.1. Example 50. Perspective directions for the development of Information Technology

Artificial Intelligence has been buzzing around for the past few years. Business experts suggest that it would be trending in every industry, including health care, manufacturing, retail, and mobile apps. AI is doing miracles when we talk about technology, whether it be mobile apps, web development, Virtual or Augmented Reality, and IoT. The majority of companies are opting for Artificial Intelligence Development to automate their tasks, retain users, drive more engagement, and reduce human workload.[9]

1. Artificial Intelligence
2. Blockchain
3. Augmented Reality and Virtual Reality
4. Cloud Computing
5. Angular and React
6. DevOps
7. Internet of Things (IoT)
8. Intelligent Apps (I—Apps)
9. Big Data
10. RPA (Robotic Process Automation)
11. 5G technology
12. Drone Technology
13. Quantum computing.

In computer science, **artificial intelligence (AI)**, sometimes called machine intelligence, is intelligence demonstrated by machines.

Artificial intelligence can be classified into three different types of systems: analytical, human-inspired, and humanized artificial intelligence.

---

[9] https://www.iqvis.com/blog/ai-powered-mobile-apps/.

Analytical AI has only characteristics consistent with cognitive intelligence; generating a cognitive representation of the world and using learning based on past experience to inform future decisions. Human-inspired AI has elements from cognitive and emotional intelligence; understanding human emotions, in addition to cognitive elements, and considering them in their decision-making.

Humanized AI shows characteristics of all types of competencies (i.e., cognitive, emotional, and social intelligence), is able to be self-conscious, and is self-aware in interactions.

In the twenty-first century, AI techniques have experienced a resurgence following concurrent advances in computer power, large amounts of data, and theoretical understanding; and AI techniques have become an essential part of the technology industry, helping to solve many challenging problems in computer science, software engineering, and operations research.[10]

A **blockchain** is a growing list of records, called blocks, that are linked using cryptography. Each block contains a cryptographic hash of the previous block, a timestamp, and transaction data (generally represented as a Merkle tree).

By design, a blockchain is resistant to modification of the data. For use as a distributed ledger, a blockchain is typically managed by a peer-to-peer network collectively adhering to a protocol for inter-node communication and validating new blocks. Once recorded, the data in any given block cannot be altered retroactively without the alteration of all subsequent blocks, which requires the consensus of the network majority. Although blockchain records are not unalterable, blockchains may be considered secure by design and exemplify a distributed computing system with high Byzantine fault tolerance.[11]

Augmented Reality and **Virtual Reality**, the twin technologies that let you experience things in virtual, that are extremely close to real, are today being used by businesses of all sizes and shapes. However, the underlying technology can be quite complex.

Cloud Computing is one of the most trending technologies.

Angular and **React** are JavaScript-based Frameworks for creating modern web applications.

Using React and Angular one can create a highly modular web app. So, you don't need to go through a lot of changes in your code base to add a new feature.

DevOps is a set of software development practices that combine software development (Dev) and information-technology operations (Ops) to shorten the systems-development life cycle while delivering features, fixes, and updates frequently in close alignment with business objectives.[12]

DevOps is a methodology that ensures that both development and operations go hand in hand. DevOps cycle is picturized as an infinite loop representing the integration of developers and operation teams by:

---

[10] **Artificial intelligence**—Wikipedia https://en.wikipedia.org/wiki/Artificial_intelligence.

[11] Wikipedia.

[12] **DevOps**—Wikipedia https://en.wikipedia.org/wiki/DevOps.

– automating infrastructure
– workflows
– continuously measuring application performance.

The **Internet of Things**, or **IoT**, is a system of interrelated computing devices, mechanical and digital machines, objects, animals or people that are provided with unique identifiers (UIDs) and the ability to transfer data over a network without requiring human-to-human or human-to-computer interaction.

In the consumer market, IoT technology is synonymous with products pertaining to the concept of the "smart home".[13] Now this concept has been expanded and is called "Smart City" and in the future "Smart Planet".

Intelligent Apps (I—Apps). I-Apps are pieces of software written for mobile devices based on artificial intelligence and machine learning technology, aimed at making everyday tasks easier.

This involves tasks like organizing and prioritizing emails, scheduling meetings, logging interactions, content, etc. Some familiar examples of I-Apps are Chatbots and virtual assistants.[14]

Big data is a field that treats ways to analyze, systematically extract information from, or otherwise deal with data sets that are too large or complex to be dealt with by traditional data-processing application software.

Data with many cases (rows) offer greater statistical power, while data with higher complexity (more attributes or columns) may lead to a higher false discovery rate Big data challenges include capturing data, data storage, data analysis, search, sharing, transfer, visualization, querying, updating, information privacy and data source.

Big data was originally associated with three key concepts: volume, variety, and velocity.

When we handle big data, we may not sample but simply observe and track what happens.

Therefore, big data often includes data with sizes that exceed the capacity of traditional software to process within an acceptable time and value.[15]

Robotic process automation (RPA) is an emerging form of business process automation technology based on the notion of metaphorical software robots or artificial intelligence (AI) workers.

In RPA systems develop the action list by watching the user perform that task in the application's graphical user interface (GUI), and then perform the automation by repeating those tasks directly in the GUI. This can lower the barrier to the use of automation in products that might not otherwise feature APIs for this purpose.[16]

---

[13] **Internet of Things** (IoT)—Wikipedia.

[14] http://cplacademy.ca/2019/03/how-to-improve-your-communication-skill/.

[15] Big Data – Wikipedia https://en.wikipedia.org/wiki/Big_data.

[16] Robotic process automation—Wikipwdia. https://en.wikipedia.org/wiki/Robotic_process_aut omation.

5G technology is going to improve processing speeds by more than 10 times in 2019. This is the technology that can make possible, for instance, the much-expected remote surgery in rural areas.[17]

Drone Technology

A drone, in technological terms, is an unmanned aircraft. Drones are more formally known as unmanned aerial vehicles (UAVs) or unmanned aircraft systems (UASes). Essentially, a drone is a flying robot that can be remotely controlled or fly autonomously through software-controlled flight plans in their embedded systems, working in conjunction with onboard sensors and GPS.

In the recent past, UAVs were most often associated with the military, where they were used initially for anti-aircraft target practice, intelligence gathering, and then, more controversially, as weapons platforms. Drones are now also used in a wide range of civilian roles ranging from search and rescue, surveillance, traffic monitoring, weather monitoring, and firefighting, to personal drones and business drone-based photography, as well as videography, agriculture, and even delivery services.[18]

Drones are one component of IoT technology.

Quantum computing is the use of quantum–mechanical phenomena such as superposition and entanglement to perform computation. A quantum computer is used to perform such computations, which can be implemented theoretically or physically.

There are currently two main approaches to physically implementing a quantum computer: analog and digital. Analog approaches are further divided into quantum simulation, quantum annealing, and adiabatic quantum computation. Digital quantum computers use quantum logic gates to do computation. Both approaches use quantum bits or qubits.

Qubits are fundamental to quantum computing and are somewhat analogous to bits in a classical computer. Qubits can be in a 1 or 0 quantum state. But they can also be in a superposition of the 1 and 0 states. However, when qubits are measured the result is always either a 0 or a 1; the probabilities of the two outcomes depend on the quantum state they were in.

Today's physical quantum computers are very noisy and quantum error correction is a burgeoning field of research. As of April 2019, no large scalable quantum hardware has been demonstrated, nor have commercially useful algorithms been published for today's small, noisy quantum computers.

4.2.   Example 51. Cortana

After the launch of Windows 10, Cortana is a well-known application to all. The app, which was available on Windows phones, is now available on Android phones as well.

---

[17] https://interestingengineering.com/5-technology-trends-to-watch-in-2019.
[18] https://internetofthingsagenda.techtarget.com/definition/drone.

It helps you manage tasks that would otherwise need hands to do. For instance, you can schedule a meeting and the rest will be handled by Cortana. Moreover, it can be used to send emails, search for desired products, or anything you can find on the internet. However, to use Cortana, you need to sign in to your Microsoft account and make use of it.[19]

4.3. Example 52. Hound

Hound is yet another application that you would love to use. Similar to Google Voice Search, you can use Hound by speaking naturally and get the results displayed instantly. Without using your finger or tapping the screen, you can activate the app by saying "Ok Hound".

Hound comes in a wide range of features such as you can listen to your favorite music or video playback curated in the Sound Hound playlist. Apart from that, you can set multiple alarms simultaneously and set the timer according to your requirements.

Similarly, you can use Hound to know the current weather or weather forecast for the coming days. Ask Hound to assist you in finding the nearest best food restaurants, find movie show time, book an Uber ride, make calculations, or search the web for any keyword to get information. More is yet to come on Hound, so make sure to stay tuned for the unique AI application.[20]

4.4. Example 53. Recent News

Recent News is a news aggregation app, powered by Artificial Intelligence algorithms that will study your reading habits. It will keep you updated on the latest articles, news, and relevant reading stuff according to your interests and past behavior. Recent News is the best application to stay informed and updated on subjects you love.

It can be synchronized with other devices to help you bookmark articles on your smartphone and browse them later on other smart devices. Moreover, the ease of exporting your history and bookmarks will also help you conserve battery by nullifying background activity.

Recent News is a must-have app designed for those who are always on the hunt for new articles, news, and blogs relevant to their field of interest.[21]

4.5. Example 54. Fyle

Artificial intelligence empowered applications have come as a surprisingly positive development. Fyle is actually one such application. Accessible for iOS clients, this application gives you a chance to track, scan, and upload all your financial costs. Called intelligent cost marketing, this application comes controlled with artificial intelligence, monitors the investments, and fundamentally, builds an organization's overall performance. The platform includes an AI-powered engine for auto-extraction of costs from Gmail and Outlook, alongside mileage tracking and

---

[19] https://www.iqvis.com/blog/top-7-best-artificial-intelligence-apps-for-android-ios/.

[20] https://www.iqvis.com/blog/top-7-best-artificial-intelligence-apps-for-android-ios/.

[21] https://www.iqvis.com/blog/top-7-best-artificial-intelligence-apps-for-android-ios/.

receipt catch by means of mobile application. Fyle can automate complex endorsement and processing work processes for brisk reimbursement, permitting a paperless, powerful framework that coordinates with accounting software. Features incorporate boundless one-click cost reports, real-time compliance and policy check, charge card reconciliation, and much more.[22]

4.6.   Example 55. Virtual Assistant DataBot: RobotBot Studio

RobotBot Studio presents you with the best artificial insight application on Android/iPhone 2019 and it is your own artificial intelligence. This application is the most fun virtual assistant application for Android. This application will reply with its voice to your demand upon any point you are keen on. This application is free and a smart digital assistant can rapidly recognize the subjects required. With this application, you can make customized media presentations utilizing voice, content, and pictures. This virtual assistant application will respond to explicit inquiries with the assistance of Google, Wikipedia, channel RSS, and online material. Here you can likewise make summarizing pages that contain your answer, related subtleties, materials, search services, web links, and so forth. The application will let you effectively share answer utilizing SMS, email, and social media platforms. It is cross-platform and can utilize a similar partner on your cell phone, tablet, and PC. This assistant is absolutely accessible utilizing voice search as Jarvis. You likewise have an alternative to customize your own assistant according to your choice and preferences and modify languages, voice, name, behavior, and can create your very own bot as Jarvis.[23]

4.7.   Example 56. ELSA

Intended to help better the English of individuals who can't speak the language, English Language Speaking Assistant, ELSA, is an application that is accessible for Android mobile phones. Generally utilized by individuals who originate from areas where English isn't the main language, ELSA is very useful for those needing to learn English. This application is intended to enable users to learn and improve their English learning and pronunciation aptitudes in a range of about a month! In addition to the fact that it helps you gain proficiency with the language, it additionally gives you an undeniable report that indicates what your advancement has been like since you began utilizing the application.[24]

4.8.   Example 57. Stifr Magic Cleaner

Stifr Magic Cleaner is also one of the popular artificial intelligence apps, used to detect junk photos on your phone to free up some space. The images are scrutinized through a series of machine learning algorithms. Apart from that, the app is quick

---

[22] https://www.analyticsinsight.net/top-5-artificial-intelligence-apps-of-2019/.
     https://www.fylehq.com/.

[23] https://www.analyticsinsight.net/top-5-artificial-intelligence-apps-of-2019/.
     http://www.databot-app.com/.

[24] https://www.analyticsinsight.net/top-5-artificial-intelligence-apps-of-2019/.
     https://elsaspeak.com/home.

and delivers results instantly. Moreover, you can select the photos manually to delete or allow the Stifr to do the work. The app is available free of any charge on iOS and Play Store.[25]

4.9.   Example 58. Google Allo

It is regarded as one of the best and most well-known artificial intelligence applications for Android and iOS users. It enables the user to complete an activity from the voice and is the best messaging application. The user can likewise use this application as a voice-to-text application on a cell phone. It additionally has an enormous collection of stickers, emoticons, and a stunning smart answer feature.[26]

4.10.   Example 59. Jarvis Artificial Intelligent

GS Tech presents you with another incredible artificial intelligence application. This AI application will make your mobile phone hands-free. It is the best digital friend which will help you at anyplace with its cool and one-of-a-kind features. It has unique voice recognition and hot word identification functions and can make a cool experience while using it in the open air and traveling. You have an alternative to prepare your new personal assistant with your own directions and reaction. The customized answer can make more fun and its online database contains all the more day-by-day updated commands. This application has numerous features with this application, you can make calls, you can set alerts, and can open any applications. The application additionally enables you to play music, open Wi-Fi, Bluetooth, and streak light, and can check time, date, and battery level. It will help you in reading messages and will give you a quick alert.[27]

4.11.   Example 60. Robin

Robin is another popular Artificial Intelligence app that works as a personal assistant while you are on the move. Easy to use, Robin allows you to write text through voice, get local information, jokes, and GPS navigation without moving your eyes off the road.

Whether you are looking for a parking place, traffic status, weather forecast, or Twitter News, the app allows you to do as much as you can think of. You can even call the person by calling the name of the person without jotting down the contact list. It is a must-have app for those who are on the move and don't want to risk their lives by looking at the screen instead of the road.[28]

---

[25] https://www.iqvis.com/blog/top-7-best-artificial-intelligence-apps-for-android-ios/.

[26] https://www.analyticsinsight.net/top-5-artificial-intelligence-apps-of-2019/.
https://play.google.com/store/apps/details?id=com.google.android.apps.fireball&hl=en_IN.

[27] https://www.analyticsinsight.net/top-5-artificial-intelligence-apps-of-2019/.
https://play.google.com/store/apps/details?id=com.pa.gs.ai&hl=en_IN.

[28] https://www.iqvis.com/blog/top-7-best-artificial-intelligence-apps-for-android-ios/.

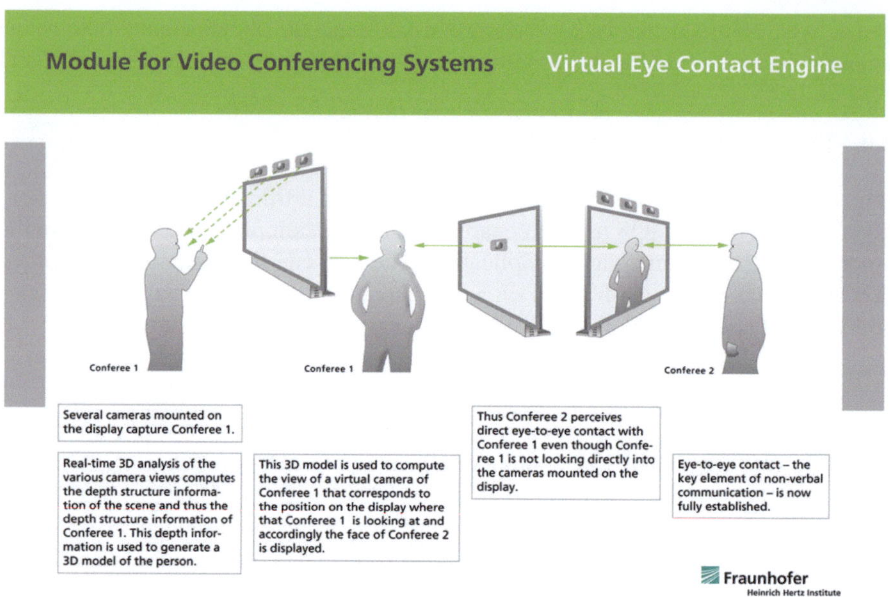

Fig. 9.5   Virtual eye contact engine

4.12.   Example 61. Virtual Eye Contact Engine

Fraunhofer Heinrich-Hertz-Institute presented the Virtual Eye Contact Engine for the first time to the public at 3D Stereo Media 2010. This prototype demonstrator allows the user to perceive direct eye contact with a remote party although the user is captured by cameras mounted on top of the display. The eye contact problem is one of the key drawbacks of current video conferencing systems and is caused by the fact that the cameras cannot be mounted at a position where the user is looking, i.e., the center of the display. By applying a real-time 3D analysis on multiple camera views and a succeeding virtual view rendering, it is possible to create a novel view of a camera virtually placed at the center of the display. The perception of eye contact awareness strongly depends on the quality of the depth estimation and on the rendering of a novel view in high quality and high resolution. This is convincingly offered by the presented system.

Several components of the Virtual Eye Contact Engine are based on research performed in the European research project FP7 3DPresence, Proposal no.: FP7-215269 (Fig. 9.5).[29]

4.13.   Example 62. Quadcopters external navigation

The Flying Machine Arena (FMA) is a portable space devoted to autonomous flight. Measuring up to $10 \times 10 \times 10$ m, it consists of a high-precision motion

---

[29] Virtual Eye Contact Engine https://www.hhi.fraunhofer.de/en/departments/vit/technologies-and-solutions/capture/virtual-eye-contact-engine.html.

Fig. 9.6 Flying machine arena

capture system, a wireless communication network, and custom software executing sophisticated algorithms for estimation and control.

The motion capture system can locate multiple objects in the space at rates exceeding 200 frames per second. While this may seem extremely fast, the objects in the space can move at speeds in excess of 10 m/s, resulting in displacements of over 5 cm between successive snapshots. This information is fused with other data and models of the system dynamics to predict the state of the objects in the future.

The system uses this knowledge to determine what commands the vehicles should execute next to achieve their desired behavior, such as performing high-speed flips, balancing objects, building structures, or engaging in a game of paddle-ball. Then, via wireless links, the system sends the commands to the vehicles, which execute them with the aid of on-board computers and sensors such as rate gyros and accelerometers.

Although various objects can fly in the FMA, the machine of choice is the quad-copter due to its agility, mechanical simplicity and robustness, and its ability to hover. Furthermore, the quadcopter is a great platform for research in adaptation and learning: it has well-understood, low-order first-principal models near hover, but is difficult to characterize when performing high-speed maneuvers due to complex aerodynamic effects. We cope with the difficult-to-model effects with algorithms that use first-principle models to roughly determine what a vehicle should do to perform a given task, and then learn and adapt based on flight data (Fig. 9.6).[30]

4.14. Example 63. New e-skin

While birds are able to naturally perceive the Earth's magnetic field and use it for orientation, humans have so far not come close to replicating this feat—at least, until now. Researchers at the Helmholtz-Zentrum Dresden-Rossendorf (HZDR) in

---

[30] https://www.flyingmachinearena.ethz.ch/.

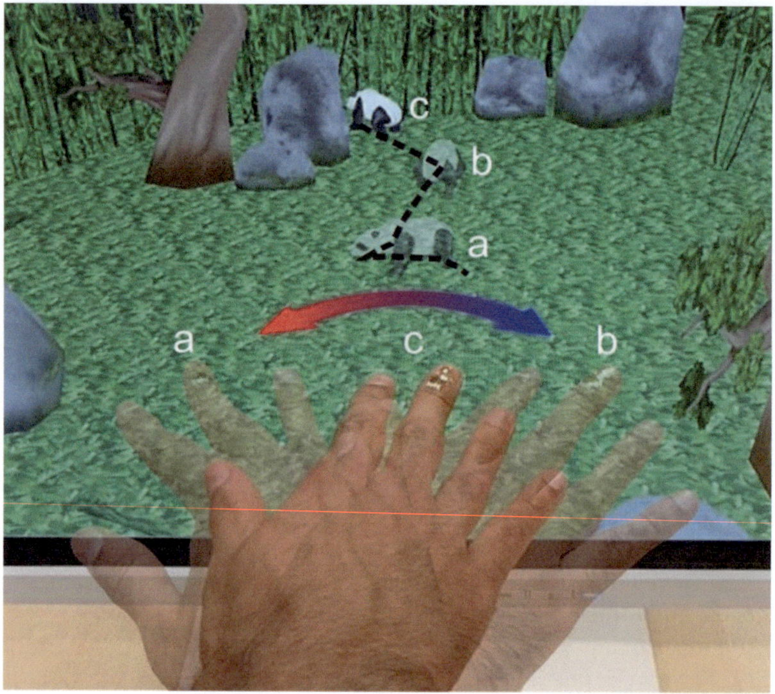

Fig. 9.7 Electronic skin points the way north

Germany have developed an electronic skin (e-skin) with magneto-sensitive capabilities, sensitive enough to detect and digitize body motion in the Earth's magnetic field. As this e-skin is extremely thin and malleable, it can easily be affixed to human skin to create a bionic analog of a compass. This might not only help people with orientation issues, but also facilitate interaction with objects in virtual and augmented reality (Fig. 9.7).[31]

4.15.   Example 64. The three waves of digitalization in healthcare

Health information and communication technology has become an essential part of a doctor's work—and it is a strategic tool for change. "We are on the brink of the third wave of digitalization of healthcare which will make medicine more precise," says Jarmo Reponen MD, PhD, Professor of Health Information Systems at the Faculty of Medicine, University of Oulu, Finland.

Professor Reponen, what is the status quo of the eHealth professionals' tools?

Researching the software tools professionals have been using in healthcare information and communication since the middle of this millennium in Finland and

---

[31] Electronic Skin points the Way North https://www.hzdr.de/db/Cms?pOid=57205&pNid=0&pLang=en.

also in collaboration with our Nordic partners, we can see three different waves of digitalization in healthcare.

The first wave supplied us with electronic medical records (EMR), digital image archives (PACS), and networks in imaging and laboratory aspects. So that all the medical data is transformed and available in digital format in hospitals and primary care institutions. In Finland and in other Nordic countries, those tools have reached practically 100% availability already 10 years ago. For example, we don't have paper records or film images anymore.

Today, we are in the middle of the second wave of digitalization in healthcare which has given us more connectivity through national health information exchanges (HIE) and tools for interacting with patients and citizens. The core of the second wave is the patient empowerment. These are essential for healthcare reform with freedom of choice and patient mobility.

While the second wave is going on, we are in the first steps into the third wave of digitalization which will empower the patients and citizens even more. It is based on the first two waves and will add artificial intelligence to professional tools through, for example, machine learning. Within the medical record systems, picture archives, and communication systems, we will see more and more intelligent tools and different support systems. For example, in many medical even today, you can have instant warnings of interactions of different medicines which are not compatible with each other.

What, if any effects, will those developments have on the patients?

Health information systems are entering the second and third wave digitalization, which will give the professionals more connectivity, more interactions with citizens, and more intelligent tools. This requires a new generation of software but also awareness of the new roles of physicians and nurses. Citizens' own role in their health will increase with the new eHealth and mHealth tools.

What advances in intelligent computing will have the greatest effect on the healthcare system?

There are plenty of effects because so many things are changing. Some of those changes will make the workflow more effective. The use of certain treatment-orientated software will make the treatment more precise. Combined they will hopefully make hospitals work more efficiently. On the other hand, artificial intelligences is coming to certain areas as a helping tool. It will help the professionals in the differential diagnosis to personalize the treatment of certain patients. The medicine is becoming more precise.

How will those changes transform diagnosis, treatment, and workflow?

We get more information about the patient to the point-of-care. With the availability of the archives of previous data, patient data from other areas of the health information exchanges, physicians can make better diagnoses. It also means that they can make more precise diagnoses and more accurate decisions on patient treatment.

Of course, it is not enough that the systems store and deliver medical data to support the efficiency of medical treatment. We need more workflow-orientated software for professionals that could give us guidance in our work. So that if we have a certain kind of patient in our consultation, the software automatically lists the tools I need to diagnose and treat that patient. These kinds of software are emerging already.

When we talk about the various intelligent and augmented tools, I would say that the vendors should look more for standardization in their software. They need to have application programming interfaces (API) and connectivity in their add-on tools because there will be always new kinds of inventions. Those vendors who can rapidly react to those innovations have the edge. I really am looking forward to software vendors that can have standardized interfaces for these news kinds of tools to be connected to the core software. That is my message to the healthcare industry.

You say that you want more decision-support tools. Don't you think there is a probability that physicians will rely too much on those tools and not enough on their own brain?

Of course, using those decision support tools will have implications on the education and teaching area. We must educate a new generation of physicians and nurses who are aware of those tools and can utilize them to see what is possible or not possible with these tools. I come from a university and therefore like to emphasize the importance of education and teaching these subjects. Most younger professionals use social media channels like YouTube of Facebook. But that doesn't help them in the professional usage of the new decision support or other eHealth tools. That is my message to universities and nursing schools: make those tools a part of your curriculums so that the new generation of professionals can better save patients in the future.[32]

4.16.   Example 65. Future Trends of AI-based Smart Systems and Services

Smart systems and services aim to facilitate growing urban populations and their prospects of virtual-real social behaviors, gig economies, factory automation, knowledge-based workforce, integrated societies, modern living, among many more. To satisfy these objectives, smart systems and services must comprises a complex set of features such as security, ease of use and user-friendliness, manageability, scalability, adaptivity, intelligent behavior, and personalization. Recently, artificial intelligence (AI) has been realized as a data-driven technology to provide an efficient knowledge representation, semantic modeling, and can support a cognitive behavior aspect of the system. In this paper, an integration of AI with smart systems and services is presented to mitigate the existing challenges. Several novel types of research work in terms of frameworks, architectures, paradigms, and algorithms are discussed to provide possible solutions against the existing challenges in AI-based smart systems and services. Such novel research works involve efficient shape image

---

[32] https://medicalview.org/the-three-waves-of-digitalization-in-healthcare/.

retrieval, speech signal processing, dynamic thermal rating, advanced persistent threat tactics, user authentication, and so on.[33]

4.17. Example 66. Smart homes and the 'Internet of Ears'

For the past couple of decades, homes have been getting steadily smarter. The next generation of smart homes may offer what researchers are calling an "Internet of Ears."

Smart homes today feature appliances, security cameras, and entertainment systems. They also feature heating and cooling systems that are interconnected and online. Online, in this context, means connected to the Internet.

We refer to the technology of interconnecting devices, such as home appliances, vehicles, etc. as the "Internet of Things" or "IoT." The technology of interconnecting government, industrial, or commercial buildings is also part of the IoT. Someday, even entire communities will be interconnected.

Smart homes of tomorrow will be buildings that adjust to people's activity with just a few, small sensors hidden in the walls. There might also be sensors under the floor. There would be no need for invasive cameras. This technology, say Ming-Chun Huang and Soumyajit Mandal, is still maybe a decade or so away.

There is actually a constant 60 Hz electrical field all around us, and because people are somewhat conductive, they short out the field just a little.

The first advantage will be energy efficiency for buildings, especially in lighting and heating, as the systems adjust to how humans are moving from one room to another, allocating energy more efficiently.

The system would also be able to track and measure a building's safety and structural integrity based on human occupancy. This could be critical in, for example, a hurricane or earthquake (Fig. 9.8).[34]

4.18. Example 67. Smart City

Smart city is emerging as a new alternative to solve various urban issues such as urban aging, traffic congestion, energy shortage, environmental pollution, and crime. Smart city is the key to discovering converged new industries that will dominate the 4th industrial revolution through data, network, and artificial intelligence (AI). Recently, there has been an increase in the number of attempts to solve urban problems using network technology both domestically and internationally. The smart city market is emerging as an innovative growth engine centered on energy, transportation, and security by utilizing information and communication technology (ICT) such as AI, big data, 5G, and networks. According to a UN report, population growth in urban

---

[33] Daewon Lee, Jong Hyuk Park. Future Trends of AI-based Smart Systems and Services: Challenges, Opportunities, and Solutions. JIPS. Volume: 15, No: 4, Page: 717–723, Year: 2019 http://jips-k.org/q.jips?cp=pp&pn=682.

[34] *"An 'Internet of Ears' for crowd-aware smart buildings based on sparse sensor networks,"* Xinyao Tang, Ming-Chun Huang, and Soumyajit Mandal. https://doi.org/10.1109/ICSENS.2017.8234263}. Conference: 2017 IEEE SENSORS.

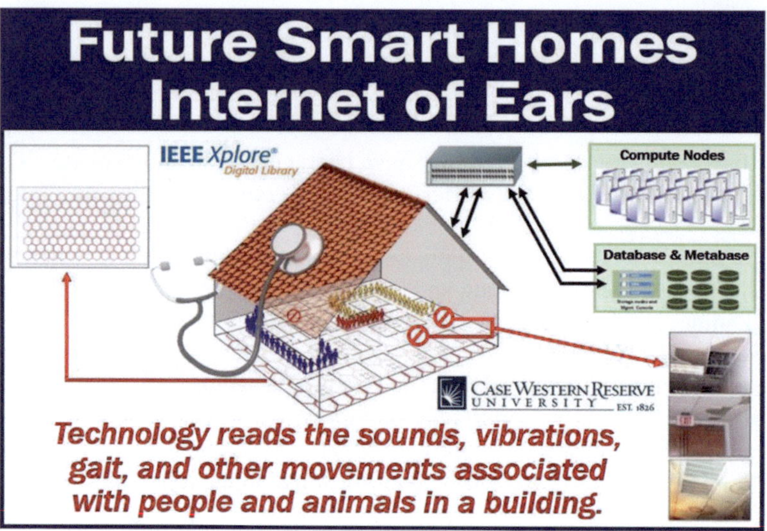

Fig. 9.8 Smart homes

areas is expected to reach 66% by 2050, with 70% of the world's resources consumed in cities.

Markets & Markets is expected to grow from $0.4 trillion in 2004 to $1.1 trillion in 2020, whereas Frost & Sullivan expects $1 trillion in 16 years and $1.5 trillion by 2020. The overall size of the smart cities market is forecast to reach $ 2.57 trillion by 2025; thus, recording a strong CAGR of 18.4% over the forecast period according to a new report from Grand View Research. In other words, smart city is an effective way of supporting economic growth by applying new technologies to control climate change and improve the quality of life for urban citizens. In addition, smart city focuses on security and sustainable, efficient control of available resources to improve economic and social performance. In the big data environment, the Internet has grown rapidly in terms of heterogeneous data consisting of complex media content such as text, images, video, audio, and graphics. These heterogeneous data are provided in various media, and they have various characteristics. Such heterogeneous data are made up of a mixture of cross-media data content and provided in a variety of sources and in various structured and unstructured formats. Data complexity, size, diversity, and uncertainty make it difficult to analyze and build models using traditional machine learning approaches. These situations need to address issues, challenges, and countermeasures for various papers and related research to meet heterogeneous computing requirements.[35]

---

[35] Jeong Hoon Jo, Pradip Kumar Sharma, Jose Costa Sapalo Sicato and Jong Hyuk Park. Emerging Technologies for Sustainable Smart City Network Security: Issues, Challenges, and Countermeasures. JIPS. Volume: 15, No: 4, Page: 765–784, 2019. http://jips-k.org/q.jips?cp=pp&pn=686.

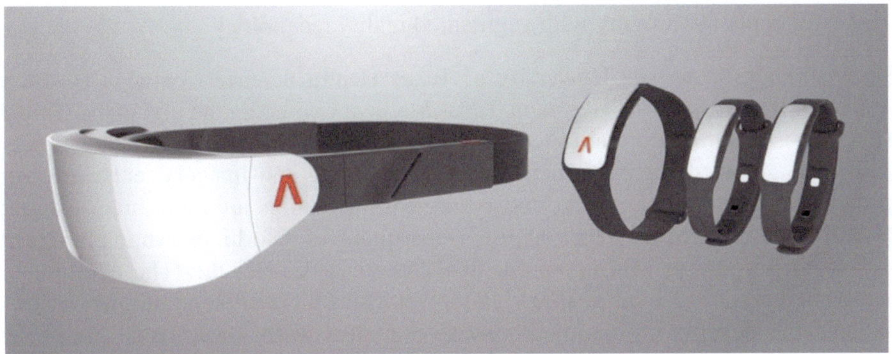

Fig. 9.9 A potential production version of SysteMD

4.19. Example 68. A sensor for tracking mental health

Mental illness, while often challenging, is very common. According to the National Institutes of Health, one in five adults experiences some form of mental health issue in any given year in the United States. The National Alliance on Mental Health reports that mental health problems are so common that one in four Americans will experience a mental health problem sometime during their lifetime.

Given these numbers, it's clear that many people will eventually encounter a mental health challenge. And yet, diagnosing and treating mental illness remains stubbornly imprecise and relies primarily on a clinician's observations and a patient's self-reporting. As a result, patients seeking treatment can get stuck in a "treat and repeat" process, sometimes for years before making therapeutic progress. "Clinicians rely largely on patient self-reporting to diagnose psychiatric disorders, guide treatment and track progress. It is an approach that can be subjective and vulnerable to bias, inaccuracy and incompleteness, potentially hindering the effectiveness of care," said Andrea Webb, PhD, a trained psychophysiologist and a principal scientist at Draper. Webb, who develops tools for diagnosing mental health disorders as a member of Draper's information and cognition division, believes that speeding up the path to developing a treatment plan and better understanding a patient's response to treatment is vital to increasing its efficiency and effectiveness and improving patient outcomes.

Webb and her team set out to develop a wearable device to complement current clinical practice. Backed by decades of research, the team at Draper knew there would be value in monitoring a patient's heart rate, skin sweating, respiration, and pupil diameter, and analyzing the data with a set of algorithms to identify patterns that could help clinical care providers with screening, diagnosing and treatment monitoring (Fig. 9.9).[36]

---

[36] https://medicalview.org/systemd-A-sensor-for-tracking-mental-health/.

4.20.   Example 69. Surgery with augmented reality technology

Sinus surgeons with The University of Texas Health Science Center at Houston (UTHealth) and Memorial Hermann-Texas Medical Center are the first in the United States to use augmented reality technology during minimally invasive sinus procedures. "Augmented reality, which uses 3-D mapping and imagery, enhances our understanding of complex anatomy so surgical procedures are more precise," said Martin J. Citardi, M.D., chair of the Department of Otorhinolaryngology-Head and Neck Surgery in McGovern Medical School at UTHealth. "The addition of augmented reality to a surgical navigation serves as a GPS-like system and offers patients the benefits of minimally invasive surgery with lower risks and better outcomes."

The initial uses of the technology included a March 2 case at Memorial Hermann-TMC in which Citardi performed revision image-guided functional endoscopic sinus surgery for recurrent chronic rhinosinusitis. In addition, the patient also had a fibro-osseous lesion blocking drainage from the left frontal sinus. "This was a complicated case. By using this technology, we were able to plan a pathway to drain that blocked frontal sinus and avoid the need for a more extensive procedure," Citardi said. He believes augmented reality technology has the potential to improve many types of sinus procedures, including those performed for chronic rhinosinusitis, sinonasal polyps, and even tumors.

The surgery was performed using Stryker's Scopis Target Guided Surgery (TGS) technology, a system designed to give surgeons the tools to plan pathways and critical structures in preoperative medical imaging scans. During surgery, this planning is overlaid with the surgeon's endoscopic view of the surgical area. The system assists the surgeon in following the defined pathway and avoiding critical structures. "This system also allows easy recording of both the surgery and surgical planning. Such digital content will be important for the training of surgeons in the difficult area of endoscopic sinus surgery. This will ultimately benefit patients," said Citardi, who is affiliated with UT Physicians, the clinical practice of McGovern Medical School, and Memorial Hermann-TMC (Fig. 9.10).

4.21.   Example 70. Blockchain may improve the medications of the future

Big data. Machine Learning. Internet of Things. Blockchain. Futuristic concepts from the world of technology will likely soon find their way into your medicine cabinet—and onto your mobile phone.

Using a prototype app for smartphones, researchers from the University of Copenhagen have taken the next step in the dosing, production, and distribution of the pharmaceutical products of the future. And the time for innovation is more than ripe, says Professor Jukka Rantanen of the Department of Pharmacy: 200 years ago, the first patent on making tablets was filed and the products have not changed much since. We are still having the same tablets. What we are doing now is suggesting a totally new type of product', he says. "By rethinking the product design principles, related manufacturing solutions and distribution models for the pharmaceutical products, it

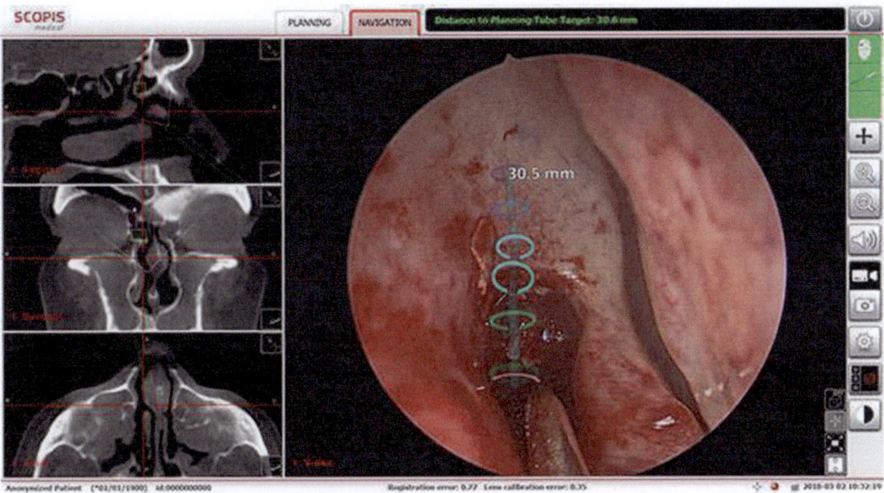

Fig. 9.10 Surgery with augmented reality technology

is possible to dramatically reduce the overall price of medicine while also improving the safety and efficacy of the medication."

The core of Jukka Rantanen and his research group's wager for a future solution for pharmaceutical products is the new concept of cryptopharmaceuticals, embodying the mentioned prototype of an app for smartphones. The app is called "MedBlockChain" and has been developed by the group's former MSc-student Lasse Nørfeldt. It is, among other things, based on the research group's earlier work on the digitalization of pharmaceutical products, for example in the form of printing medications as edible QR codes.

With the app, patients will be able to scan a medication and receive confirmation that it is a genuine product and not a fake item. A problem that, according to Jukka Rantanen, is particularly serious in countries with less structured medicines regulatory agencies.

At the same time, patients can choose to provide access to a range of personal data—everything from heart rate monitor watches, pedometers, and internet-connected bath scales to genetic profiles, screen time, and social media usage—all contributing to knowledge that can enable computer systems based on artificial intelligence to gradually pin down the optimal dose for each patient.

"This type of data already exists in our information-rich society. It would be logical to employ this big data for something useful. Not just for sharing on Facebook, your exercise app or something like that, but also for defining your optimal dose of given medicine," says Jukka Rantanen.

With the growing mass of personal data, data security is also gaining importance, Jukka Rantanen points out. To guarantee data security, the app uses the so-called blockchain technology, which is probably best known in connection with the cryptocurrency Bitcoin. With blockchain, information—or data blocks—are linked in a chain that cannot be changed without simultaneously altering all other links of information in the chain. Thus, all changes will be detected and may be traced. If something looks suspicious, the system can also generate an alarm.

As an example, a patient who scans a QR code on his medication may be alerted by an alarm if the code does not match the one that the pharmaceutical company has entered into the system, or if the medication does not match the prescription. Conversely, the pharmaceutical company may be alerted if an otherwise unique medication code is registered more than once.

Likewise, an absence of registrations may form the basis for alarms as it may reveal that the patient is not taking his or her medication as planned. This information may for example be shared with the patient's doctor or relatives.

The blockchain concept may still seem distant to most people, but in fact, the technology is already being used in similar ways for everything from insurance and finance to shipping and food, explains Jukka Rantanen.

As an example, Chinese consumers have already become accustomed to scanning items in the supermarket to confirm that the product they are buying is, for example, indeed bacon produced in Denmark, and not a counterfeit product.' All of this is technologically possible. Now, the big question is how we should handle all of this data and who should get access to it. That is the discussion we hope to start with this new concept of cryptopharmaceuticals', says Jukka Rantanen.

He emphasizes Denmark as an obvious candidate as a pioneer country for technology. Among other things based on the country's existing tradition of storing citizens' health data and the prominent pharmaceutical industry. "I think it has huge potential for Denmark to be among the first movers on this type of product. It is not limited to only one clinical condition. There could be a completely new type of product family coming out of this," says Jukka Rantanen.

For the research team, the next step is to test the app on a test group of patients. This could for example be diabetes, where patients are most often accustomed to taking medication and measuring their personal blood sugar on a regular basis (Fig. 9.11).[37]

4.22.   Example 71. Using mHealth for detecting atrial fibrillation

A new study highlights the feasible use of mobile health (mHealth) devices to help with the screening and detection of atrial fibrillation (AF), a common heart condition.

AF is the most common heart rhythm disturbance, affecting around one million people in the UK. People with AF are at increased risk of having a stroke and dying, as well as heart failure and dementia. Currently, low detection due to lack of visible symptoms and nonadherence are major problems in current management approaches for patients with suspected AF.

mHealth devices, such as fitness trackers, smartwatches, and mobile phones, may enable earlier AF detection, and improved AF management through the use of photoplethysmography (PPG) technology. PPG is a simple and low-cost optical technique that can be used to detect blood volume changes in the microvascular bed of tissue. It is often used non-invasively to make measurements at the skin surface.

---

[37] https://medicalview.org/blockchain-may-improve-the-medications-of-the-future/.

Fig. 9.11 Blockchain may improve the medications

Researchers aimed to determine the feasibility of AF screening in a large population-based cohort using smart devices with PPG technology, combined with a clinical care AF management pathway. As part of the study, AF screening was performed with smart wristbands or watches using PPG technology made available for the population aged over 18 years across China for approximately seven months.

Overall 187,912 participants used smart devices to monitor their pulse rhythm. During this time 424 (0.23%) of the individuals received a "suspected AF" notification. Of those 227 (87%) were confirmed as having AF by health providers and other secondary examinations. These patients were provided with therapy and successfully anticoagulated.

Based on the study, continuous home-monitoring with smart device-based PPG technology could be a feasible, cost-effective approach for AF screening. There were 95% of patients following entry into a program of integrated AF care, and approximately 80% of high-risk patients were successfully anticoagulated. This would help efforts at screening and detection of AF, as well as early interventions to reduce stroke and other AF-related complications (Fig. 9.12).

4.23.   Example 72. Using a virtual 3D heart to fight atrial fibrillation

A team at the University of Auckland 's Bioengineering Institute has created a virtual 3D heart that could have a major impact on the treatment of the most common heart rhythm disturbance, atrial fibrillation (AF).

Dr. Jichao Zhao and his international team have developed a very "intelligent" machine learning algorithm, alias AtriaNet, that can simultaneously utilize local and

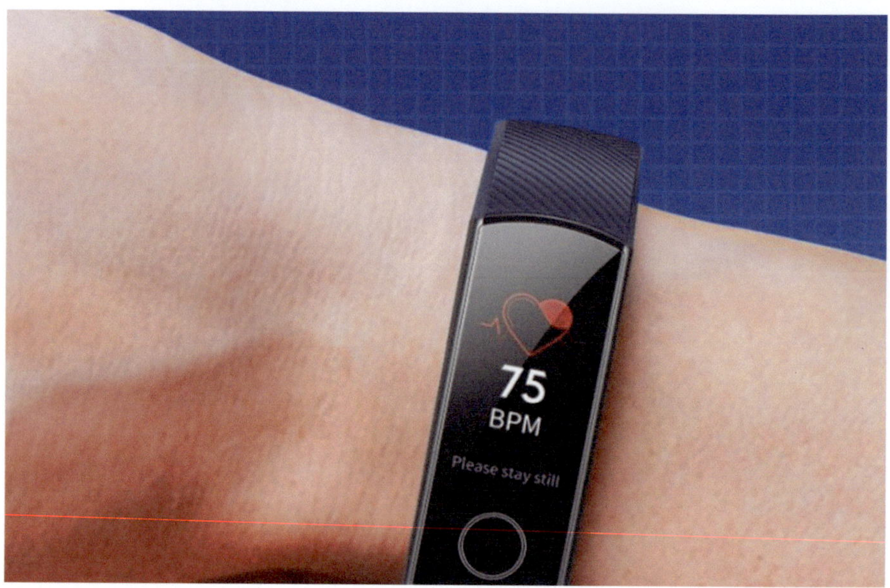

Fig. 9.12 Using mHealth for detecting atrial fibrillation

global information of whole-body MRI scans to accurately reconstruct and visualize atrial geometry in 3D. AtriaNet was developed using 154 three-dimensional, contrast-enhanced MRIs, (the largest dataset of its kind in the world), to separate atrial chambers from the rest of the human body. AtriaNet can be run on a personal computer and produce a 3D virtual heart within one minute from any new patient MRI with a high accuracy of 94%.

This advanced machine learning algorithm is way faster and more accurate than any other approach. It will help doctors to pinpoint precise locations of diseased tissue in the upper chambers (atria) of the heart by creating a 3D virtual heart within a minute.

Current clinical treatment is unsatisfactory mainly due to a lack of understanding of the human atrial tissue which directly sustains AF. This is a very important step towards much improved clinical diagnosis, patient stratification, and clinical guidance during ablation treatment for patients with atrial fibrillation (Fig. 9.13).[38]

4.24. Example 73. Augmented realty helps to save lives in war zones

Researchers at Purdue University use augmented reality tools to connect healthcare professionals in war zones, natural disasters, and in rural areas to perform complicated procedures with more experienced surgeons and physicians around the world. "The most critical challenge is to provide surgical expertise into the battlefield when it is most required," said Juan Wachs, Purdue's James A. and Sharon M. Tompkins

---

[38] https://medicalview.org/using-A-virtual-3d-heart-to-fight-atrial-fibrillation/.

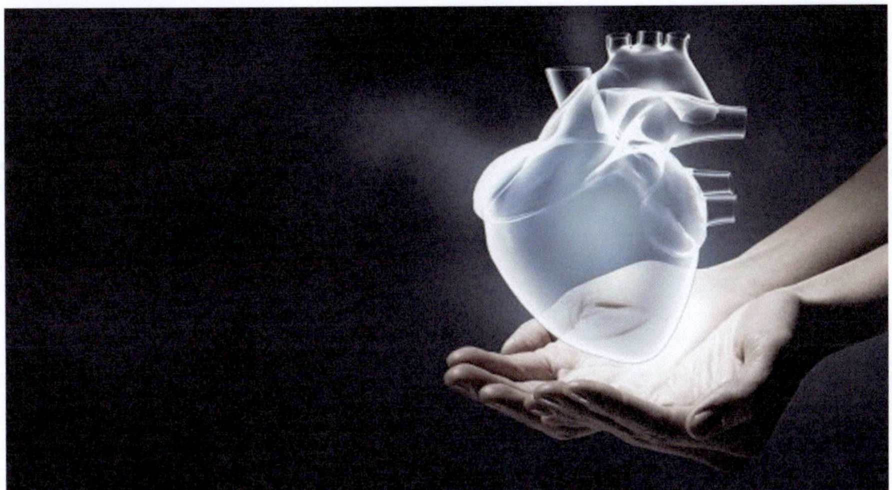

Fig. 9.13 Using a virtual 3D heart to fight atrial fibrillation

Rising Star Associate Professor of Industrial Engineering, who led the project team. "Even without having highly experienced medical leaders physically co-located in the field, with this technology we can help minimize the number of casualties while maximizing treatment at the point of injury."

The AR headset worn by the mentee in the field is designed to replace current telestrator technology, which uses a separate video screen and freehand sketches to provide feedback. "There is an unmet need for technology that connects health care mentees in rural areas with experienced mentors," said Edgar Rojas Muñoz, a doctoral student in industrial engineering, who worked on the project. "The current use of a telestrator in these situations is inefficient because they require the mentee to focus on a separate screen, fail to show upcoming steps and give the mentor an incomplete picture of the ongoing procedure."

The Purdue system features a transparent headset screen display that allows the mentee to see the patient in front of them, along with real-time on-screen feedback from the mentor. That mentor is at a separate location using a video monitor to see the AR feed and provide instant feedback to the field surgeon.

Purdue's system uses computer vision algorithms to track and align the virtual notes and marks from the mentor with the surgical region in front of the mentee. "Our technology allows trainees to remain focused on the surgical procedure and reduces the potential for errors during surgery," Muñoz said.

The U.S. Department of Defense supported the research as it looks to connect its medical professionals out in the field with specialists back at the bases who can provide critical guidance during procedures.

The Purdue technology has gone through a round of clinical evaluation and will soon go through another one (Fig. 9.14).

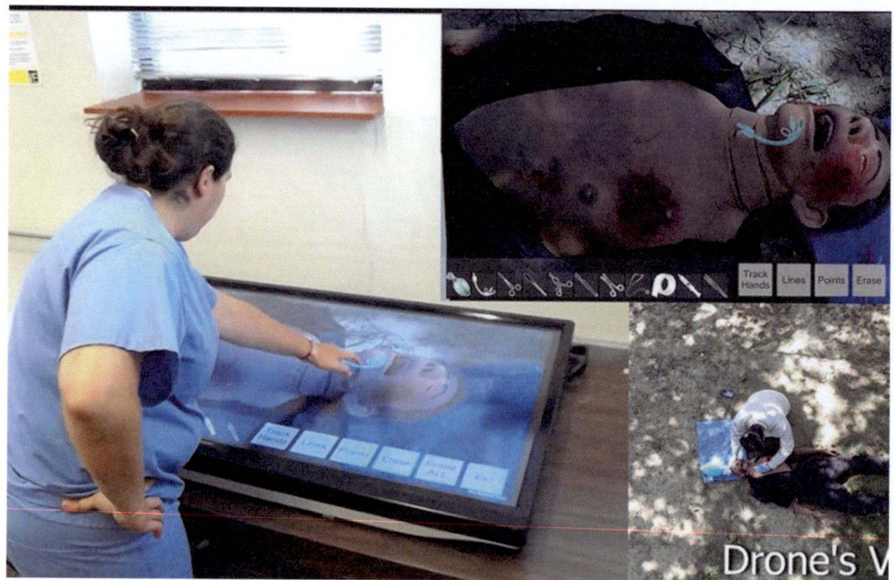

Fig. 9.14 Augmented realty helps to save lives in war zones

4.25. Example 74. Augmented reality as low aid

Nearly one in 30 Americans over the age of 40 experience low vision—a significant visual impairment that can't be corrected with glasses, contact lenses, medication, or surgery. In a new study of patients with retinitis pigmentosa, an inherited degenerative eye disease that results in poor vision, Keck School of Medicine of USC researchers found that adapted augmented reality (AR) glasses can improve patients' mobility by 50% and grasp performance by 70%. "Current wearable low vision technologies using virtual reality are limited and can be difficult to use or require patients to undergo extensive training," says Mark Humayun, MD, PhD, director of the USC Dr. Allen and Charlotte Ginsburg Institute for Biomedical Therapeutics, codirector of the USC Gayle and Edward Roski Eye Institute and University Professor of Ophthalmology at the Keck School. "Using a different approach—employing assistive technology to enhance, not replace, natural senses—our team adapted AR glasses that project bright colors onto patients' retinas, corresponding to nearby obstacles."

Patients with retinitis pigmentosa wore adapted AR glasses as they navigated through an obstacle course based on an FDA-validated functional test. Using video of each test, researchers recorded the number of times patients collided with obstacles, as well as the time taken to complete the course.

Patients averaged 50% fewer collisions with the adapted AR glasses. Patients also were asked to grasp a wooden peg against a black background—located behind four other wooden pegs—without touching the front items. Patients demonstrated a 70% increase in grasp performance with the AR glasses. "Patients with retinitis pigmentosa have decreased peripheral vision and trouble seeing in low light, which

Fig. 9.15 Augmented reality as low aid

makes it difficult to identify obstacles and grasp objects. They often require mobility aids to navigate, especially in dark environments," says Anastasios N. Angelopoulos, study project lead in Humayun's research laboratory at the Keck School. "Through the use of AR, we aim to improve the quality of life for low vision patients by increasing their confidence in performing basic tasks, ultimately allowing them to live more independent lives (Fig. 9.15)."[39]

4.26.   Example 75. Diabetes data analysis to improve insulin delivery

The lives of people with Type 1 diabetes could be significantly enhanced through algorithms that connect glucose monitors and insulin pumps to automatically regulate blood glucose to healthy levels, in the same fashion.

A new project aims to use artificial intelligence and big data techniques to analyze information gathered from thousands of continuous glucose monitors and insulin pumps. Researchers will use that information to improve algorithms that control these critical devices.

People with Type 1 diabetes must test their blood sugar often to decide how much insulin they should inject using a needle or insulin pump. Until fairly recently, blood sugar could only be tested by performing a finger stick to obtain a blood sample that would be analyzed by a glucose meter—a process most people only do 5–6 times a day.

Now, continuous glucose monitors can give people a better idea of whether their blood sugar is trending high or low by providing blood sugar estimates every five minutes, without frequent finger sticks (Fig. 9.16).[40]

---

[39] https://medicalview.org/augmented-reality-glasses-as-low-vision-aid/.

[40] https://medicalview.org/diabetes-data-analysis-to-improve-insulin-delivery/.

Fig. 9.16 Diabetes data analysis to improve insulin delivery

4.27.   Example 76. Virtual reality helps get rid of post-traumatic stress disorder

Post-traumatic memory stress disorders.

   Third-person memory helps a person to forget about them. To this end, Associate Professor of Psychology at the University of Alberta Peggy St Jacques used Virtual Reality.

4.28.   Example 77. Prediction of schizophrenia

The researchers, led by Bo Cao from the University of Alberta's (U of A) Department of Psychiatry, used the machine learning algorithm to examine functional magnetic resonance imaging (fMRI) images of both newly diagnosed, previously untreated schizophrenia patients and healthy subjects. By measuring the connections of the superior temporal cortex to other regions of the brain, the algorithm successfully identified patients with schizophrenia at 78 percent accuracy. In addition, the algorithm also predicted with 82 percent accuracy whether or not a patient would respond positively to a specific antipsychotic treatment named risperidone. "This is the first step, but ultimately we hope to find reliable biomarkers that can predict schizophrenia before the symptoms show up," said Cao, an assistant professor of psychiatry at the U of A. "In the future, with the help of machine learning, if the doctor can select the best medicine or procedure for a specific patient at the first visit, it would be a good step forward."

5. *Define the kind of Su-field*

5.1. Example 78. Optogenetics

Until now, researchers looking to stimulate specific neurons had to rely on bursts of electricity—an imprecise and difficult-to-control technique. That's why the new field of optogenetics is so exciting. By combining fiberoptics and designer viruses, researchers can now stimulate neurons with a high degree of precision. This could allow, for example, the development of implants that can take over the functions of a brain region that might have been damaged by a wound or stroke. First, the brain is injected with a virus that is engineered to activate specific neurons when light hits them. A fiber-optic cable combined with an electrode then sends light into the brain, turning the neurons on and off, on command. Initial experiments used rodents, but researchers have now applied the technique to monkeys, and DARPA recently announced a project aimed at using optogenetics to help injured veterans (Fig. 9.17).[41]

5.2. Example 79. Mechanophores

America's infrastructure needs renewal, but we can't just rebuild everything at once: We need effective ways to figure out which structures are closest to failure. One approach is to integrate tiny wireless sensors into new construction. Another is to incorporate "mechanophores," a class of materials recently developed at the University of Illinois that change color when they are stressed. Mechanophores could give an engineer a quick visual indication of whether a bridge is at risk and where the trouble lies. The researchers are currently working to tune the reaction so that it can

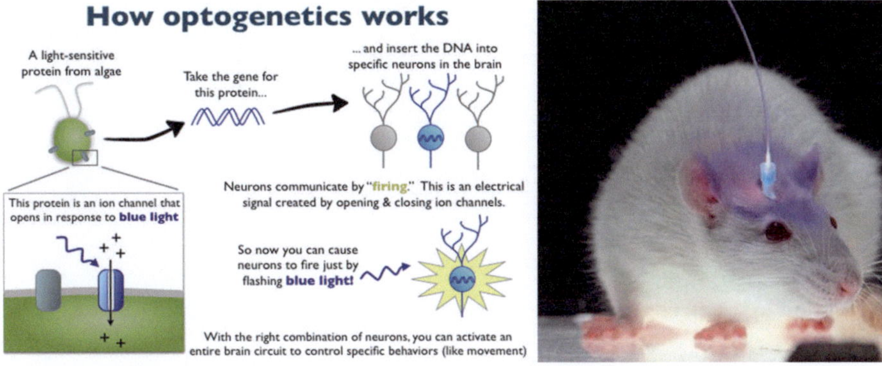

Fig. 9.17 Optogenetics[42]

[41] https://www.popularmechanics.com/technology/a6422/10-tech-concepts-you-need-to-know-for-2011/.

[42] https://futurism.com/scientists-turn-brain-cells-on-and-off-using-light-receive-high-honors.

occur at any desired level of stress. They also hope to develop new mechanophores that undergo a self-healing response when they are damaged.[43]

5.3.   Example 80. Complex-Event Processing

Corporations and governments routinely comb through enormous databases of information and images (such as those pulled from surveillance cameras) in search of patterns. But in today's data-rich world, an unfavorable signal-to-noise ratio can make it time-consuming and expensive to find anything relevant. A new generation of software is shifting the focus from "data" (a record of what's happened) to "events" (what's happening right now). Companies like StreamBase Systems and Tibco offer complex-event processing systems that analyze enormous flows of data in real time using new databases and pattern-recognition approaches. This allows them to make instant decisions about whether to make a stock trade, initiate surveillance on a potential terrorist, or halt a suspicious credit-card transaction. As the technology matures, we can expect these capabilities to trickle down to consumer devices. This would allow, for example, a GPS-enabled cellphone to sift through a constant stream of location-aware offers and alert users only to ones they would actually be interested in—such as deals on coffee along their morning commute route during the hours when they make the trek (Fig. 9.18).[44]

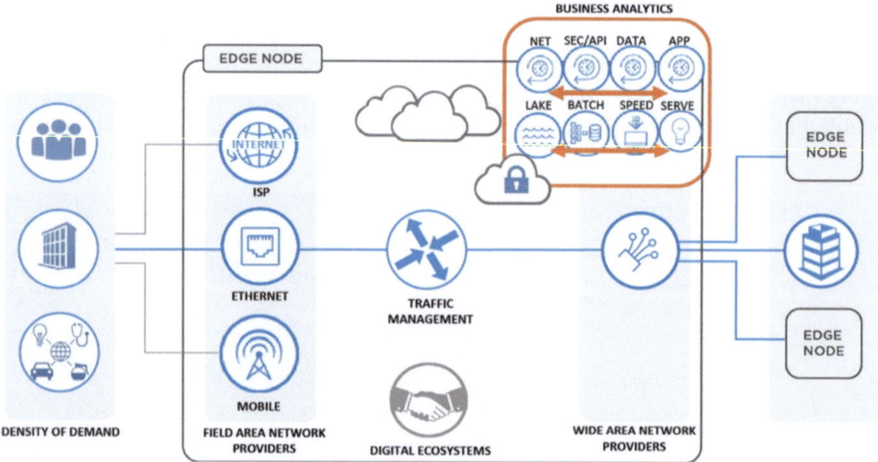

Fig. 9.18 Complex event processing design pattern[45]

---

[43] https://www.popularmechanics.com/technology/a6422/10-tech-concepts-you-need-to-know-for-2011.

[44] https://www.popularmechanics.com/technology/a6422/10-tech-concepts-you-need-to-know-for-2011.

[45] https://blog.equinix.com/blog/2018/03/28/developing-A-successful-end-to-end-complex-event-processing-strategy/.

5.4. Example 81. Homomorphic Encryption

Researchers at IBM recently cracked a decades-old problem: how to encrypt data so that other people can sort and search it without actually revealing the contents. As cloud computing becomes more pervasive over the next year, this "homomorphic" encryption will allow companies to store sensitive data on remote servers, where it can be kept secret from the server's host, but still be easily accessed and searched. Users will also be able to enter search-engine queries and receive results without the search engine ever knowing or having a record of their query. The key breakthrough was a "double-blind" scheme that can check for encryption errors and fix them without revealing the data. Best of all, the researchers demonstrated that the technique can be implemented in just a few minutes on a standard PC, not just high-priced super-computers.

5.5. Example 82. Cellphone Diagnostics

While trained medical care is a rare commodity in the developing world, cellphones are increasingly common. In fact, between 80 and 90% of the world's population now lives within range of a cell tower. That makes phones a powerful tool for bringing modern medicine to remote and poor areas. One approach pioneered by MIT spinoffs Sana Mobile and ClickDiagnotics is to have rural health workers transmit X-rays and other medical information via cellphone to far-off experts for diagnosis. Meanwhile, scientists at the University of California, Berkeley, and a PM BreakthroughAward-winning researcher at UCLA have combined inexpensive microscope parts with off-the-shelf phones to produce devices that can record and instantly analyze microscopic images, detecting malaria parasites or tuberculosis-causing bacteria. The Berkeley-designed diagnostic tool, called CellScope, will be deployed in field trials in 2011.[46]

5.6. Example 83. Making DNA data storage a reality

With continued improvements in the volume of information that can be packed into DNA's tiny structure—data can be stored at densities well into millions of gigabytes per gram—such a future doesn't look so fanciful. As the costs of oligonucleotide synthesis and sequencing continue to fall, the challenge for researchers and companies will be to demonstrate that using DNA for storage, and maybe even other tasks currently carried out by electronic devices, is practical (Fig. 9.19).[47]

5.7. Example 84. Aircraft without moving parts

MIT engineers have built and flown the first-ever plane with no moving parts. Instead of propellers or turbines, the light aircraft is powered by an "ionic wind"—a silent but mighty flow of ions that is produced aboard the plane, and that generates enough thrust to propel the plane over a sustained, steady flight.

---

[46] https://www.popularmechanics.com/technology/a6422/10-tech-concepts-you-need-to-know-for-2011.

[47] Making DNA Data Storage a Reality. The-scientist. *Oct 1, 2017* https://www.the-scientist.com/cover-story/making-dna-data-storage-a-reality-30218.

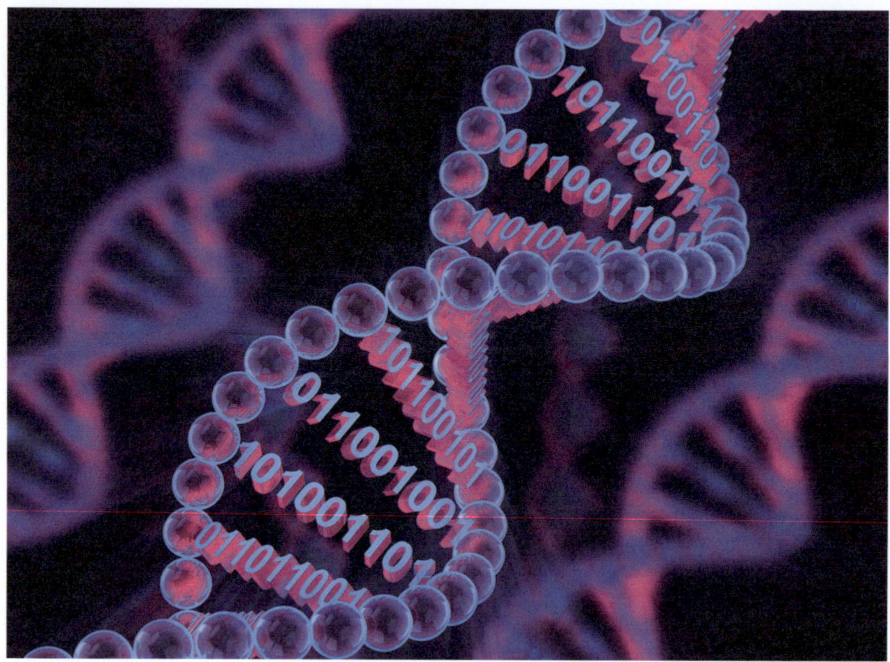

Fig. 9.19 DNA data storage gets random access[48]

"This is the first-ever sustained flight of a plane with no moving parts in the propulsion system," says Steven Barrett, associate professor of aeronautics and astronautics at MIT. "This has potentially opened new and unexplored possibilities for aircraft which are quieter, mechanically simpler, and do not emit combustion emissions."

He expects that in the near-term, such ion wind propulsion systems could be used to fly less noisy drones. Further out, he envisions ion propulsion paired with more conventional combustion systems to create more fuel-efficient, hybrid passenger planes and other large aircraft (Fig. 9.20).[49]

5.8.   Example 85. Control and identification of goods in the store

The invention relates to a product identification system when the goods are purchased by a buyer in a self-service store. The product identification system in the self-service store includes a scanning device, a goods-collecting device with an individual identification number, a camera, at least one self-service payment terminal, and a server. The scanning device reads the identification number, scans the barcodes of goods, and creates a list of goods and a resulting barcode. The camera is installed on

---

[48] https://spectrum.ieee.org/the-human-os/biomedical/devices/dna-data-storage-gets-random-access.

[49] MIT engineers fly first-ever plane with no moving parts http://news.mit.edu/2018/first-ionic-wind-plane-no-moving-parts-1121.

Fig. 9.20 Aircraft without moving parts

Fig. 9.21 Control and identification of goods in the store. US 2019 69 695 A1

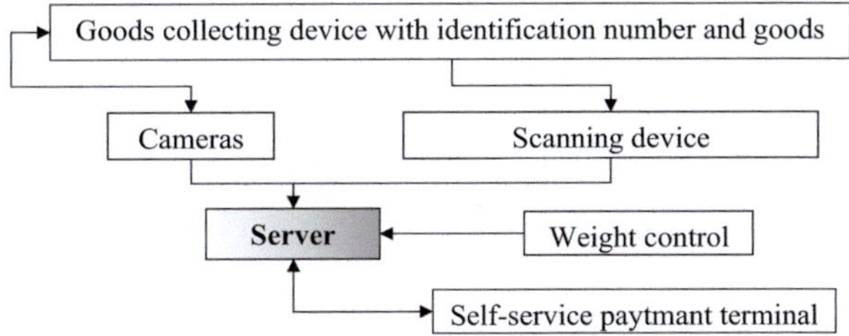

the goods collecting device and reads images of the goods and tracks their movement. The terminal sends information to the server, wherein the server receives and analyzes information from the camera and terminal. In case when the information from the camera and from the resulting barcode coincide, the buyer pays for the purchase, and receives a check (Figs. 9.11 and 9.21).[50]

5.9. Example 86. X-ray reading

A new artificial intelligence algorithm can reliably screen chest X-rays for more than a dozen types of disease, and it does so in less time than it takes to read this sentence, according to a new study led by Stanford University researchers. The algorithm, dubbed CheXNeXt, is the first to simultaneously evaluate X-rays for a

---

[50] US 2019 69 695 A1 https://patentswarm.com/patents/US20190069695A1.

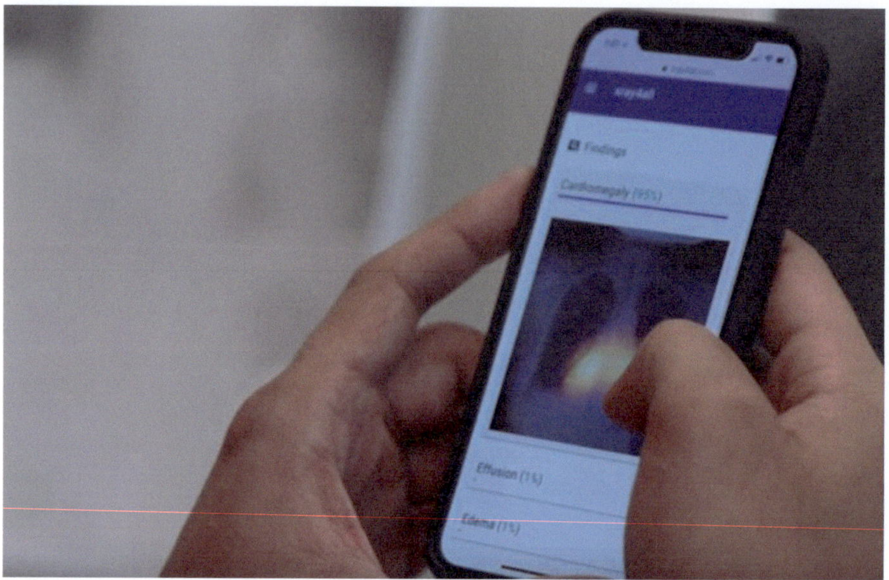

Fig. 9.22 X-ray reading

multitude of possible maladies and return results that are consistent with the readings of radiologists, the study says.

Scientists trained the algorithm to detect 14 different pathologies: For 10 diseases, the algorithm performed just as well as radiologists; for three, it under-performed compared with radiologists; and for one, the algorithm outdid the experts (Fig. 9.22).[51]

5.10.   Example 87. Patient surveillance system

The WAVE Clinical Platform is an always-on remote monitoring platform that displays near real-time clinical views of physiologic and medically relevant data including waveforms and alarms for at-risk patients across hospital workstations, mobile devices, and inside electronic medical records.

WAVE automatically calculates risk, giving an at-a-glance early warning of patient deterioration up to six hours in advance of when clinicians would otherwise notice—and while there is still time to prevent further deterioration.

In fact, while using the Visensia Safety Index, WAVE's first FDA-cleared predictive algorithm, UPMC went from six unexpected deaths in their control group to zero (Fig. 9.23).[52]

---

[51] https://medicalview.org/chexnext-outperformed-radiologists-in-evaluating-chest-X-rays/.

[52] Excel Medical's WAVE Clinical Platform Receives FDA Clearance https://www.cardiovascul arbusiness.com/topics/practice-management/excel-medicals-wave-clinical-platform-receives-fda-clearance.

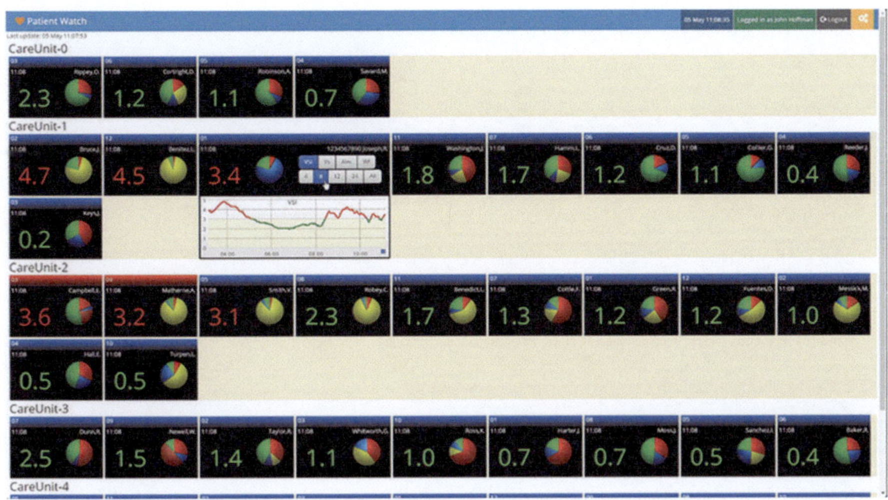

Fig. 9.23 Patient surveillance system[53]

5.11. Example 88. New technique to simultaneously target fibroblasts while killing cancer cells

Cancer cells surround fibroblasts (other cells) that protect cancer cells from the immune system. Fibroblasts also supply cancer cells with nutrients, necessary for growth.

Scientists have equipped a virus that kills carcinoma cells with a protein so it can also target and kill adjacent cells…

Scientists have equipped a virus that kills carcinoma cells with a protein so it can also target and kill adjacent cells that are tricked into shielding the cancer from the immune system.

It is the first time that cancer-associated fibroblasts within solid tumors—healthy cells that are tricked into protecting cancer from the immune system and supplying it with growth factors and nutrients—have been specifically targeted in this way.

The researchers, who were primarily funded by the Medical Research Council (MRC) and Cancer Research UK, say that if further safety testing is successful, the dual-action virus—which they have tested in human cancer samples and in mice—could be tested in humans with carcinomas as early as next year.

Currently, any therapy that kills the "tricked" fibroblast cells may also kill fibroblasts throughout the body—for example, in the bone marrow and skin—causing toxicity.

In this study, the researchers used a virus called enadenotucirev, which is already in clinical trials for treating carcinomas. It has been bred to infect only cancer cells, leaving healthy cells alone.

---

[53] https://www.medgadget.com/2018/05/the-wave-patient-surveillance-and-predictive-algorithm-platform-interview-with-mark-koppel-cmo-of-excel-medical.html.

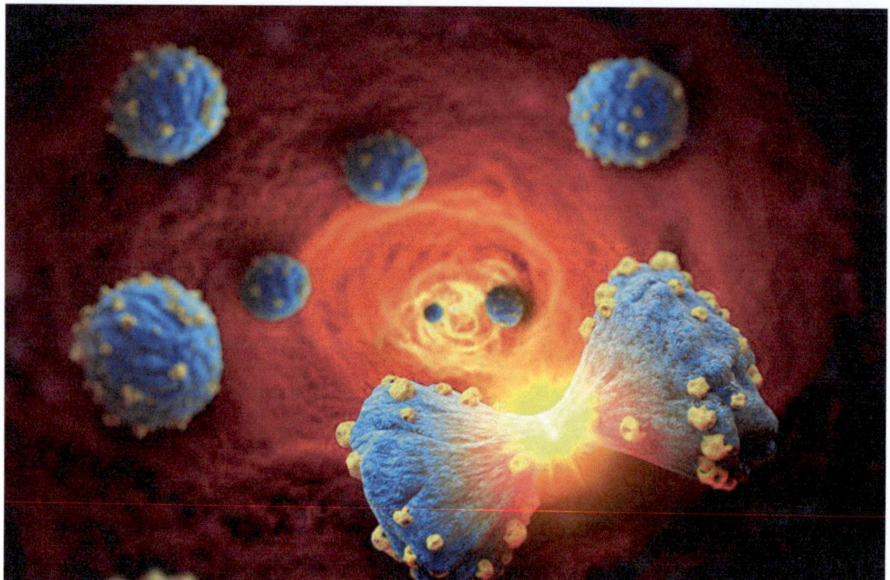

Fig. 9.24 New technique killing cancer cells

They added genetic instructions into the virus that caused infected cancer cells to produce a protein called a bispecific T-cell engager.

The protein was designed to bind to two types of cells and stick them together. In this case, one end was targeted to bind to fibroblasts. The other end specifically stuck to T cells—a type of immune cell that is responsible for killing defective cells. This triggered the T cells to kill the attached fibroblasts (Fig. 9.24).[54]

5.12. Example 89. Gait recognition

Chinese authorities have begun deploying a new surveillance tool: "gait recognition" software that uses people's body shapes and how they walk to identify them, even when their faces are hidden from cameras.

Already used by police on the streets of Beijing and Shanghai, "gait recognition" is part of a push across China to develop artificial-intelligence and data-driven surveillance that is raising concern about how far the technology will go.

Huang Yongzhen, the CEO of Watrix, said that its system can identify people from up to 50 m (165 feet) away, even with their back turned or face covered. This can fill a gap in facial recognition, which needs close-up, high-resolution images of a person's face to work.

Gait analysis can't be fooled by simply limping, walking with splayed feet, or hunching over, because we're analyzing all the features of an entire body.

---

[54] https://www.drugtargetreview.com/news/36803/killing-cancer-cells/.

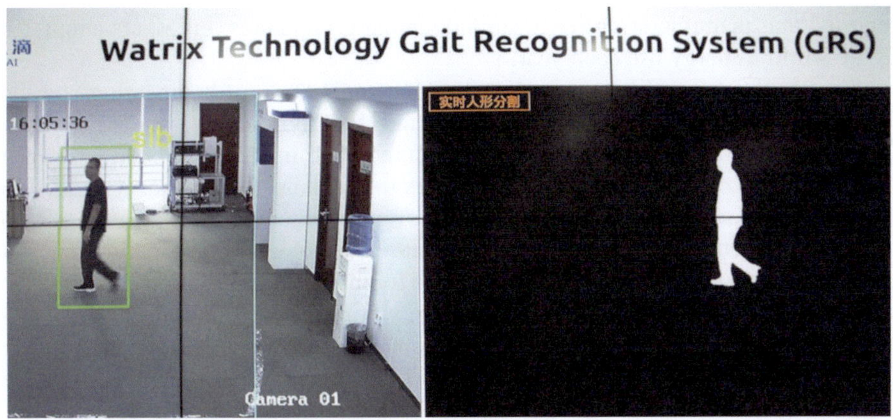

Fig. 9.25 Gait recognition

Chinese police are using facial recognition to identify people in crowds and nab jaywalkers, and are developing an integrated national system of surveillance camera data.

Watrix's software extracts a person's silhouette from a video and analyzes the silhouette's movement to create a model of the way the person walks. It isn't capable of identifying people in real-time yet. Users must upload a video into the program, which takes about 10 min to search through an hour of video. It doesn't require special cameras—the software can use footage from surveillance cameras to analyze gait.

Beyond surveillance, Huang says gait recognition can also be used to spot people in distress such as elderly individuals who have fallen down. Nixon believes that the technology can make life safer and more convenient (Fig. 9.25).[55]

5.13.   Example 90. Fertilizer energy reduction

Nitrogen-based synthetic fertilizer forms the backbone of the world food supply, but its manufacture requires a tremendous amount of energy. Now, computer modeling at Princeton University points to a method that could drastically cut the energy needed by using sunlight in the manufacturing process.

Manufacturers currently make fertilizer, pharmaceuticals, and other industrial chemicals by pulling nitrogen from the air and combining it with hydrogen. Nitrogen gas is plentiful, making up about 78 percent of air. But atmospheric nitrogen is hard to use because it is locked into pairs of atoms, called N2, and the bond between these two atoms is the second strongest in nature. Therefore, it takes a lot of energy to split up the N2 molecule and allow the nitrogen and hydrogen atoms to combine. Most manufacturers use the Haber–Bosch process, a century-old technique that exposes the N2 and hydrogen to an iron catalyst in a chamber heated to more than 400 degrees Celsius. The method uses so much energy that Science magazine recently

---

[55] https://www.apnews.com/bf75dd1c26c947b7826d270a16e2658a.

reported that manufacturing fertilizer and similar compounds represents about 2% of the world's energy use each year.

A research team led by Emily Carter, Princeton's Dean of engineering and the Gerhard R. Andlinger Professor in Energy and the Environment, wanted to know if it would be possible to use light to weaken the bond in the atmospheric nitrogen molecule. If so, it would allow manufacturers to cut radically the energy needed to split nitrogen for use in fertilizer and a wide array of other products.

"Harnessing the energy in sunlight to activate inert molecules such as nitrogen, and greenhouse gases methane and carbon dioxide for that matter, is a grand challenge for sustainable chemical production," said Carter, who is a professor of mechanical and aerospace engineering and of applied and computational mathematics. "Replacing traditional energy-intensive high temperature, high-pressure chemical manufacturing with sunlight-driven, room-temperature processes is another way to decrease our dependence on fossil fuels."

The researchers were interested in taking advantage of the unique behavior of light when it interacts with metallic nanostructures smaller than a single wavelength of light. Among other effects, the phenomenon, called surface plasmon resonance, can concentrate light and enhance electric fields. John Mark Martirez, a postdoctoral researcher and member of the Princeton research team, said that the researchers believed it would be possible to use plasmon resonances to boost a catalyst's power to split nitrogen molecules.

"It is a different method of delivering energy to break the bond," he said. "Instead of using heat, we are using light (Fig. 9.26)."[56]

5.14. Example 91. Prediction of survivability of exoplanets

Artificial intelligence is giving scientists new hope for studying the habitability of planets, according to a study from astronomers Chris Lam and David Kipping.

Their work looks at so-called "Tatooines," and uses machine learning techniques to calculate how likely such planets are to survive into stable orbits.

Circumbinary planets are those planets that orbit two stars instead of just one, much like the fictional planet Tatooine in the Star Wars franchise. Tens of these planets have so far been discovered, but working out whether they may be habitable or not can be difficult.

Moving around two stars instead of just one can lead to large changes in a planet's orbit, which means that it is often either ejected from the system entirely, or it crashes violently into one of its twin stars. Traditional approaches to calculating which of these occurs for a given planet get significantly more complicated as soon as the extra star is thrown into the mix.

---

[56] New process could slash energy demands of fertilizer, nitrogen-based chemicals https://www.pri nceton.edu/news/2018/01/17/new-process-could-slash-energy-demands-fertilizer-nitrogen-based-chemicals.

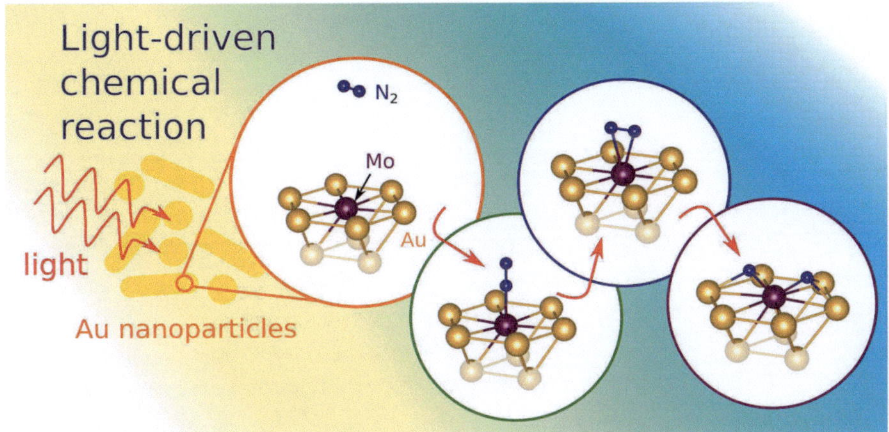

Fig. 9.26 Fertilizer energy reduction

"When we simulated millions of possible planets with different orbits using traditional methods, we found that planets were being predicted as stable that were clearly not, and vice versa," explains Lam, lead author of the study and a recent graduate of Columbia University.

Planets need to survive for billions of years in order for life to evolve, so finding out whether orbits are stable or not is an important question for habitability. The new work shows how machine learning can make accurate predictions even if the standard approach—based on Newton's laws of gravity and motion—breaks down.

"Classification with numerous complex, interconnected parameters is the perfect problem for machine learning," says Professor Kipping, supervisor of the work.

After creating ten million hypothetical Tatooines with different orbits, and simulating each one to test for stability, this huge training set was fed into the deep learning network. Within just a few hours, the network was able to out-perform the accuracy of the standard approach.

More circumbinary planets look set to be discovered by NASA's Transiting Exoplanet Survey Satellite (TESS) mission, and Lam expects their work to help: "Our model helps astronomers to know which regions are best to search for planets around binary stars. This will hopefully help us discover new exoplanets and better understand their properties (Fig. 9.27)."[57]

5.15. Example 92. Nano Targeting of Plant Nutrients

An innovative technology developed at the Technion could lead to significant increases in agricultural yields. Using a nanometric transport platform on plants that was previously utilized for targeted drug delivery, researchers increased the penetration rate of nutrients into the plants, from 1% to approximately 33%.

---

[57] AI beats astronomers in predicting survivability of exoplanets https://phys.org/news/2018-04-ai-astronomers-survivability-exoplanets.html.

Fig. 9.27 Prediction of survivability of exoplanets

The findings were recently published in Scientific Reports ("Therapeutic nanoparticles penetrate leaves and deliver nutrients to agricultural crops").

The use of nanotechnology for targeted drug delivery is a new approach, and is the focus of the research activity being conducted at the Laboratory for Targeted Drug Delivery and Personalized Medicine Technologies at the Wolfson Faculty of Chemical Engineering.

The present research, which repurposed the technology for agricultural use, was performed by the laboratory director, Assistant Professor Avi Schroeder, and graduate student, Avishai Karny.

"The constant growth in the world population demands more efficient agricultural technologies, which will produce more and healthier foods and reduce environmental damage," said Prof. Schroeder. "The present work provides a new means of delivering essential nutrients without harming the environment."

The researchers loaded the nutrients into liposomes—small spheres generated in the laboratory, comprised of a fatty outer layer enveloping the required nutrients. The liposomes are stable in the plant's aqueous environment and can penetrate into the cells.

In addition, the Technion researchers can "program" the liposomes to disintegrate and release the load at precisely the location and time of interest, namely, in the roots and leaves. Disintegration occurs in acidic environments or in response to an external signal, such as light waves or heat. Of note, the molecules comprising the particles are derived from soy plants and are therefore approved and safe for consumption by both humans and animals.

In the present experiment, the researchers used 100-nm liposomes to deliver nutrients—iron and magnesium—into both young and adult tomato crops. They demonstrated that the liposomes, which were sprayed in the form of a solution onto the leaves, penetrated the leaves and reached other leaves and roots. Only when reaching the root cells did they disintegrate and release the nutrients.

In addition to demonstrating the high effectiveness of this approach, as compared to the standard spray method, the researchers also assessed the regulatory limitations associated with the spread of volatile particles.

"Our engineered liposomes are only stable within a short spraying range, up to 2 m," explained Prof. Schroeder. "If they travel in the air beyond that distance, they break down into safe materials (phospholipids). We believe that the success of this study will expand the research and development of similar agricultural products, to increase the yield and quality of food crops and nourish a 7-billion community of people on earth (Fig. 9.28)."[58,59]

5.16.   Example 93. Smartphone fast charging

The next-gen Quick Charge standard will be even faster, while ensuring better thermal management.

True Fast Wireless Charging tech makes use of Dual Charge capability to deliver 15W power. To recall, the Dual Charge feature debuted with the latest Quick Charge 4+ standard, and it utilizes an additional power management IC in the smartphone. This allows the device to charge faster, while ensuring the thermals are kept in check. Unlike conventional charging methods, the power flows through two paths, which spreads the charge and thus reduces the power and better heat dissipation. Qualcomm claims that this enables QC4+-laden devices with Dual Charge functionality to charge up to 3 °C cooler.

Taking Dual Charge technology, a step further, the San Diego-based company is planning to launch Triple Charge capability next year. This means, even faster charging speeds, without worrying about the device running hot. As per Qualcomm, the tech will cross the current 32W power limit it has for charging. For reference, Huawei's recently-announced Mate 20 Pro comes with 40W SuperCharge, which promises to juice up to 70% levels in just 30 min. Currently, the chipmaker claims that QC4+-based smartphones can charge up to 50% within 15 min, which means we can expect the next-gen standard to be even faster, maybe even surpassing OPPO's famed Super VOOC charging, which can go from zero to 100 within just 35 min for the Find X Lamborghini Edition (Fig. 9.29).[60]

---

[58] https://www.technion.ac.il/en/2018/05/nano-targeting-of-plant-nutrients/as3/.

[59] https://www.nanowerk.com/nanotechnology-news/newsid=50367.php.

[60] https://www.91mobiles.com/hub/qualcomm-quick-charge-roadmap-details/.

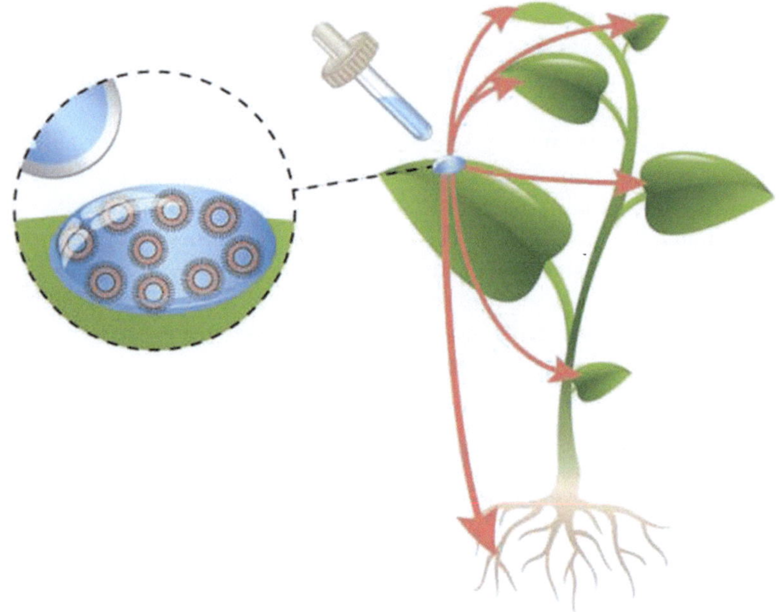

Fig. 9.28 Nano targeting of plant nutrients

Fig. 9.29 Smartphone fast charging[61]

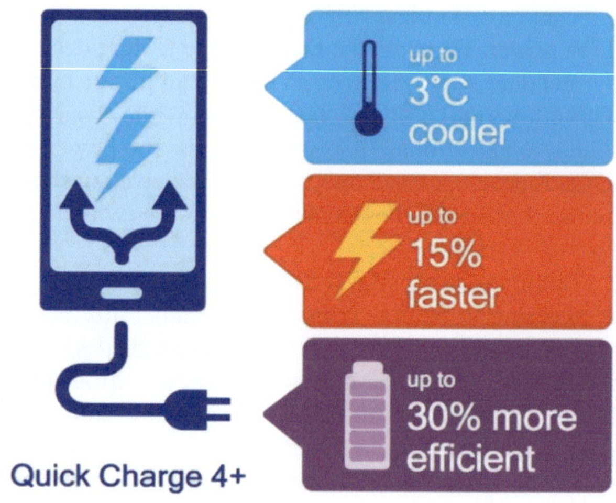

[61] https://www.hardwarezone.com.sg/tech-news-qualcomms-quick-charge-4-faster-cooler-and-more-efficient.

5.17. Example 94. Nanotechnology versus Viruses

An international interdisciplinary team of virologists and biochemists that includes scientists at Freie Universität has developed low-cost and "cell-friendly" nanogels that can efficiently prevent viral infections. The flexible nanogels mimic cell surface receptors where several viral families bind. Pathogens adhere to the nanogel molecules, so the likelihood of an infection of cells decreases significantly. The researchers in the project are based at the Potsdam Institute of Biochemistry and Biology and the Indian Association for the Cultivation of Science as well as the Institute of Chemistry and Biochemistry and the Center for Infection Medicine, both at Freie Universität. The interdisciplinary cooperation was funded through the Collaborative Research Center 765 of Freie Universität and the German Research Foundation (DFG). The research findings were published online on *ACS Nano*.

Medical research faces innumerous challenges due to the vast number of viruses that exist. Most drugs available today are effective against only a single virus or a few similar ones because they only block specific viral proteins to disrupt the viral replication cycle in the infected cells. In addition to possible side effects of these substances, the emergence of resistant virus strains represents a serious medical hazard.

Many viruses share some similarities despite their variety of species and morphology. Often, they interact multivalently with specific receptors and co-receptors on cell surfaces, which they use for initial contact and diffusion into the interior of the cells. The breakdown of the complex mechanism of cell–virus interaction is thus a key research area for developing effective broad-spectrum antiviral agents.

Heparan sulfate (HS) proteoglycans form entry ports for a variety of viruses in the cell membrane. So far, various types of nanoparticles have been designed to block viral entry via the HS molecules. They are based primarily on rigid materials such as gold and silver particles. Little research has been done on softer and more flexible substances as an alternative.

The German-Indian research team has now succeeded in developing nanogels with different degrees of flexibility that mimic cellular HS proteins. The active compound based on dendritic polyglycerol sulfate can effectively and permanently bind and shield viruses, thus preventing infections. These nanogels offer the advantage that they can flexibly adapt to the virus surface. This increases their multivalent interactions with the virus particles and reduces the likelihood that the pathogens will be able to detach again. The researchers synthesized two sulfated nanogels that work against herpes and arteriviruses in humans and other animals. The generated nanogels can achieve an inhibitory effect of up to 90 percent. The substances remain active for a relatively long time and also provide protection against virus particles released from already infected cells. Nanogels can be prepared at very low cost compared with the production of conventional antiviral drugs. Thus, they can increasingly also be used to treat animals. In addition, the polymeric gels are harmless and "cell-friendly"—unlike rigid, inflexible materials—and can be broken down into smaller fragments and excreted by the kidneys.[62]

Researchers from, among others, Chalmers University of Technology, Sweden, have now managed, for the first time, to make the surface of a gold object melt at room temperature.

5.18. Example 95. Gold melts at room temperature

To melt metals requires the application of high levels of heat. This fundamental of physics has been altered by a research team who have succeeded in melting gold at room temperature. This comes down to tension.

As tension increases strange things begin to happen at the atomic level. This is especially so with gold atoms. Scientists working at Chalmers University of Technology have succeeded in making the surface of a gold object melt at room temperature.

Gold in its purest form is a bright, slightly reddish yellow, dense, soft, malleable, and ductile metal. The melting point of gold is 1064.18 degrees Celsius. For practical use, gold is usually alloyed with base metals for use in jewelry, altering its hardness and ductility, melting point, color, and other properties.

A new means to melt gold was detected when researcher Ludvig de Knoop was examining a fragment of gold via an electron microscope. de Knoop noted that at the highest level of magnification with an increased electric field the gold atoms changed.

On closer study of recordings made by the microscope, de Knoop observed that the surface layers of gold had melted at room temperature. This was a phenomenon that had not previously been observed.

The process was undertaken by exposing the atoms of a gold cone to a strong electric field. The field excites the gold atoms, breaking their connections to each other and causing the surface layers to melt. What was happening was that the electric field was causing the gold atoms to lose their ordered structure and release almost all their connections to each other.

What else was of interest was by further tests, it was also possible to switch the gold between a solid and a molten structure.

Based on a set of theoretical calculations from the researchers, it is now possible to melt gold at room temperature based on a process called low-dimensional phase transition. This would need to be tested out on a larger scale. However this proceeds, These insights could open up new avenues in materials science (Fig. 9.30).[63,64]

---

[62] https://www.fu-berlin.de/en/presse/informationen/fup/2018/fup_18_214-antivirale-nanogele/index.html.

[63] https://www.kitco.com/news/2018-11-29/Scientists-Succeed-At-Melting-Gold-At-Room-Temperature.html.

[64] http://www.digitaljournal.com/tech-and-science/science/new-process-melts-gold-at-room-temperature/article/537875.

Fig. 9.30 This illustration shows the atoms of a gold cone exposed to a strong electric field (via Alexander Ericson)[65]

5.19. Example 96. Gravity-Based Energy Storage

The company Energy Vault has proposed a new energy-saving technology in the form of a "Gravity-based energy storage."

A new method attempts to use one of the most powerful and constant forces on the planet for energy—gravity.

Energy Vault, a company based out of California and Nevada, had just announced its first two clients based on for its gravity-centric energy storage solution—the Indian power giant Tata and the Mexican building materials company CEMEX.

The company says that its gravity towers are based on hydroelectricity, but without the need for water. Instead, they use "custom-made concrete bricks" that the company says will not degrade over time. These bricks are lifted when there's excess energy to go around, then are given a controlled drop when more energy needs to be generated. Energy Vault claims its system can deliver a capacity between 10 and 35 MWh, and that Tata has ordered a system delivering a full 35. This, obviously, does not generate power, but rather serves as a way to store it indefinitely. Lift the bricks when there is excess power, then lower them to regenerate them when need be (Fig. 9.31).[66]

5.20. Example 97. Solid-state batteries ready to debut in home appliances

Solid-state batteries have the potential to dominate the next generation of batteries, letting Japanese companies regain prominence in a field they once led.

---

[65] https://www.geek.com/news/scientists-melt-gold-at-room-temperature-1763168/.

[66] https://www.renewableenergyworld.com/articles/2018/11/gravitybased-energy-storage-hits-the-market.html#gref.

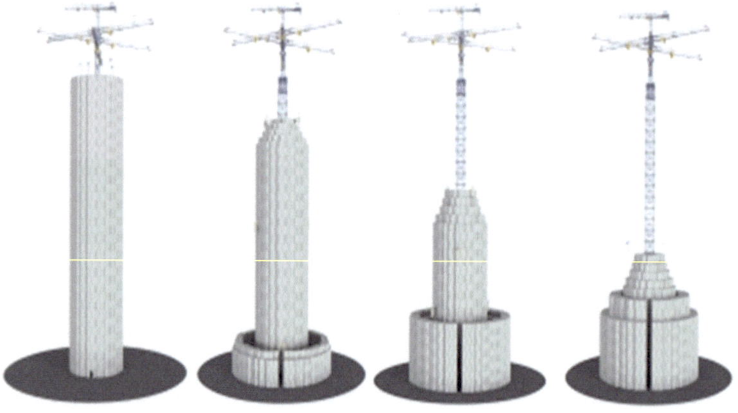

Fig. 9.31 Gravity-based energy storage[67]

Although the use of rechargeable solid-state batteries in electric vehicles has drawn the most attention, it is in consumer electronics that this new technology could make its impact felt first.

Japanese electronic components maker TDK is at the forefront with a tiny solid-state battery that can sit on a fingertip. That signals a chance for Japanese companies to reclaim market share from their rivals in South Korea and China in the highly competitive field.

Unlike the current lithium-ion generation of batteries, solid-states have no liquid that can leak so they are safer, and they also have the capacity to hold a far greater charge.

---

[67] https://www.businesswire.com/news/home/20181106006096/en/Energy-Vault-Announces-Commercial-Availability-Transformative-Utility-Scale.

Fig. 9.32 Solid-state batteries ready to debut in home appliances

TDK's solid-state battery is only a few millimeters long and can be recharged 1,000 times. The company has already begun shipping samples and is setting up operations for mass production (Fig. 9.32).[68]

5.21. Example 98. 3D scanning of the whole human body

An amazing new PET/CT[69] scanner has produced its first images of human subjects, giving scientists and clinicians new opportunities to treat cancer and other diseases. The EXPLORER is a high-sensitivity total-body positron emission tomography scanner developed by a collaboration of many different scientists. It is capable of imaging the entire human body in less than a second and with excellent fidelity.

It was originally conceived by Simon Cherry and Ramsey Badawi, two scientists at UC Davis, but years of engineering and scientific work were required to turn it

---

[68] Solid-state batteries ready to debut in home appliances https://asia.nikkei.com/Business/Bus iness-trends/Solid-state-batteries-ready-to-debut-in-home-appliances.

[69] **Positron-emission tomography (PET)** is a nuclear medicine functional imaging technique that is used to observe metabolic processes in the body as an aid to the diagnosis of disease. The system detects pairs of gamma rays emitted indirectly by a positron-emitting radioligand, most commonly fluorine-18, which is introduced into the body on a biologically active molecule called a radioactive tracer. Different ligands are used for different imaging purposes, depending on what the radiologist/ researcher wants to detect. Three-dimensional images of tracer concentration within the body are then constructed by computer analysis. In modern PET computed tomography scanners, three-dimensional imaging is often accomplished with the aid of a computed tomography X-ray scan performed on the patient during the same session, in the same machine.—Wikipedia https://en.wik ipedia.org/wiki/Positron_emission_tomography.

into reality. The first model was eventually built by United Imaging Healthcare out of Shanghai, China, and the company is working toward commercializing it and bringing it to clinics and research laboratories. The scanner's first work has been conducted at the Department of Nuclear Medicine at the Zhongshan Hospital in Shanghai.

Because of the scanner's high efficiency, it is able to produce images in as little as a second using a standard radiation dose, much faster than with conventional devices. Moreover, to help reduce radiation exposure, the dose can be reduced at the expense of just a few extra seconds of the scanner's time. And if optimal image quality is key, a longer scan at a standard dose will provide impressive results.

The developers of the device believe that new whole-body studies, which can assess how different tissues and organs react to different stimuli, will be able to be performed. The spread of inflammation, the impact of different disorders, and the mobility of cancer tumors should also be subject to easier assessment using the new scanning technology.[70]

The cylindrical EXPLORER total-body scanner combines PET and X-ray computed tomography (CT) imaging technologies to produce images with high temporal resolution.

A total of 40 rings are present in the scanner along with 48 modular block detectors. The ring diameter is approximately 80cm and the spatial resolution is 4mm. The device has an acceptance angle of 46°.

With an axial field of view (FOV) of 194 cm, the scanner provides PET images through around 500,000 detector elements.

The PET scanner is said to perform total-body scanning up to 40 times better than existing commercial scanners, performing 3D scans of the whole body within 30 s. The fast-scanning ability of the scanner avoids the need for anesthesia when scanning young children.

In addition, the unit incorporates modern solid-state silicon photomultiplier light sensors to provide high-resolution images (Fig. 9.33).[71]

5.22. Example 99. Stretchy solar cells a step closer

Organic solar cells that can be painted or printed on surfaces are increasingly efficient, and now show promise for incorporation into applications like clothing that also require them to be flexible.

The Rice University lab of chemical and biomolecular engineer Rafael Verduzco has developed flexible organic photovoltaics that could be useful where constant, low-power generation is sufficient.

Organic solar cells rely on carbon-based materials including polymers, as opposed to hard, inorganic materials like silicon, to capture sunlight and translate it into

---

[70] First Images from EXPLORER Total-Body Positron Emission Tomography Scanner https://www.medgadget.com/2018/11/first-images-from-explorer-total-body-positron-emission-tomography-scanner.html.

[71] EXPLORER Total-Body PET Scanner https://www.medicaldevice-network.com/projects/explorer-total-body-pet-scanner/.

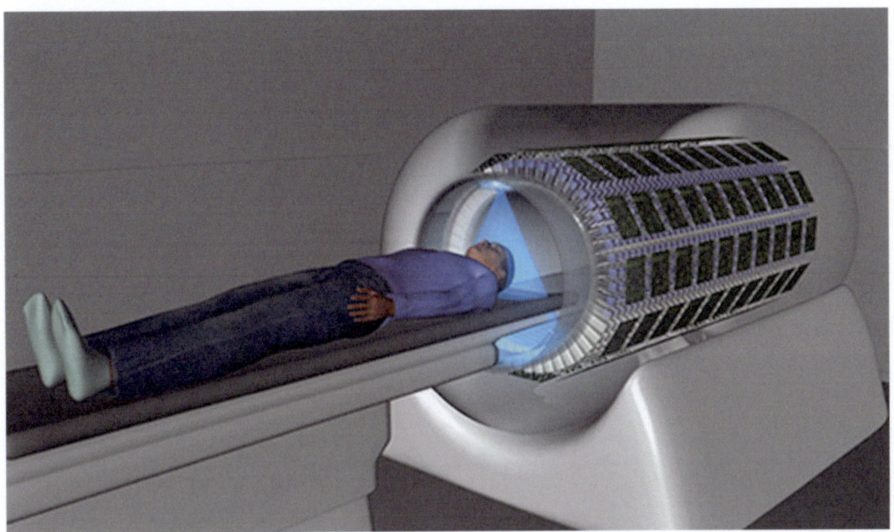

Fig. 9.33 3D scanning of the whole human body

current. Organics are also thin, lightweight, semitransparent, and inexpensive. While middle-of-the-road, commercial, silicon-based solar cells perform at about 22% efficiency—the amount of sunlight converted into electricity—organics top out at around 15% (Fig. 9.34).[72]

5.23. Example 100. Quantum "compass"

UK scientists have demonstrated the world's first standalone quantum "compass," which allows highly accurate navigation without the need for satellites.

Researchers from Imperial College London and Glasgow-based laser firm M Squared who built the device say it overcomes many of the issues of traditional GPS systems, such as blockages from tall buildings or signal jamming.

The quantum accelerometer relies on the precision and accuracy possible by measuring the properties of supercool atoms. At extremely low temperatures, the atoms behave in a "quantum" way, acting like both matter and waves.

The device represents the UK's first commercially viable quantum accelerometer, which is a far more advanced version of the sorts of accelerometers that currently exist in today's mobile phones and laptops.

The quantum accelerometer relies on the precision and accuracy made possible by measuring properties of super-cool atoms, which means any loss in accuracy is "immeasurably small (Fig. 9.35)"[73]

---

[72] Stretchy solar cells a step closer https://news.rice.edu/2018/11/07/stretchy-solar-cells-a-step-closer-2/.

[73] https://www.imperial.ac.uk/news/188973/quantum-compass-could-allow-navigation-without/.

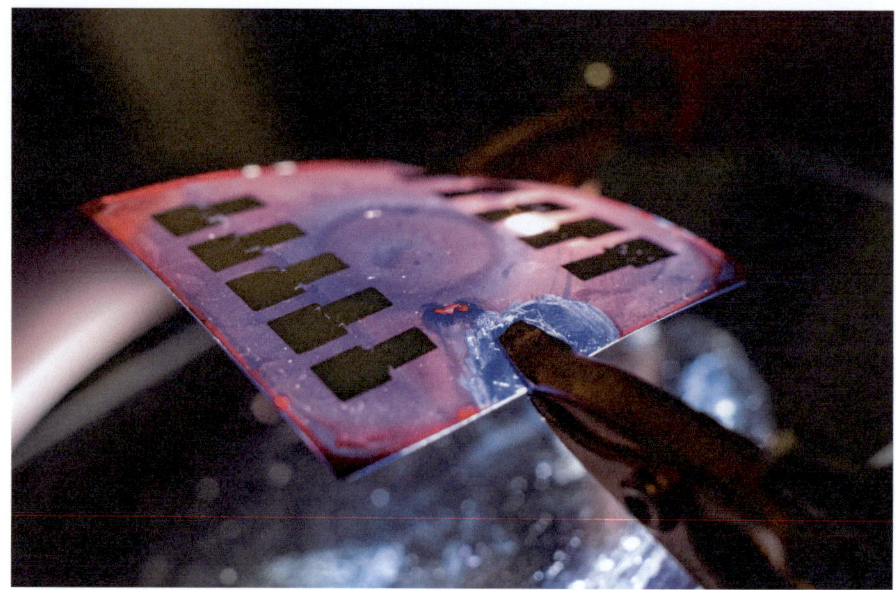

Fig. 9.34 Stretchy solar cells a step closer

Fig. 9.35 Quantum "compass"

5.24. Example 101. Liquid metal robot

A group of scientists from China and Australia designed a robot based on liquid metal, inspired by the movie "Terminator 2: Judgment Day." They hope that the development can be used for the benefit of mankind.

In a recent interview, Professor of robotics from the University of Suzhou, Li Xiangpeng, spoke about his new development.

"We were inspired by the T-1000 from Terminator 2: Judgment Day," said Xiangpeng in an interview with the South China Morning Post.

Fortunately, it was not the instinct of the robot-killer from the movie that prompted the scientist to create a liquid metal robot. On the contrary, he was interested in his ability to change shape. Despite the fact that his robot is still much inferior in its abilities to the one that James Cameron showed to the world, this is the first step towards creating advanced machines.

Xiangpeng and other researchers from China and Australia described the technology in detail in the article for Advanced Materials.

The robot consists of only three parts: a plastic wheel, a small lithium battery, and drops of gallium-based liquid metal alloy. The voltage of the battery changes the center of gravity of the liquid metal, which contributes to the flow of the robot in one direction or the other.

The researchers believe that their development will serve as inspiration for other devices—just as the film at one time prompted them to create this little robot.

"We hope to further develop soft robots using liquid metal. They can be used in special missions like search and rescue of earthquake victims, since robots can change shape to penetrate under doors or crawl through spaces that a person cannot fit into," said researcher Tan Shiyan (Fig. 9.36).[74]

5.25. Example 102. World's largest supercomputer "brain" powered up

The SpiNNaker neuromorphic supercomputer has been in planning and development for over 20 years at The University of Manchester. Today, at long last, the finished million-processor-core machine has been switched on for the first time. The project was initially backed by the Engineering and Physical Sciences Research Council (EPSRC) and is currently supported by the European Human Brain Project.

SpiNNaker is short for Spiking Neural Network Architecture computer and has a million processor cores at its disposal. This immense quantity of processor cores can model a billion biological neurons in real time and is capable of 200 trillion operations per second. The SpiNNaker neuromorphic computer can model more biological neurons in real time than any other machine on the planet.

Steve Furber, ICL Professor of Computer Engineering in the School of Computer Science at The University of Manchester, credited with the conception of the SpiNNaker project, said he was extremely excited by the potential of the computer. However, compared with real organic brains the core count vs neuron count is still

---

[74] Chinese scientists have created a liquid metal robot http://earth-chronicles.com/science/chinese-scientists-have-created-a-liquid-metal-robot.html.

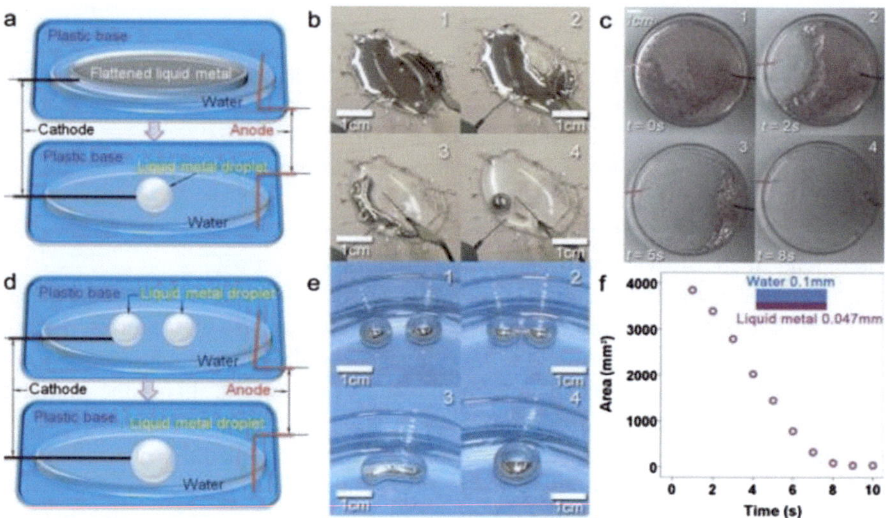

Fig. 9.36   Liquid metal robot

quite low. For example, a mouse brain consists of around 100 million neurons and the human brain is 1000 times bigger than that—and the human neurons are connected via approximately 1 quadrillion synapses.

One of the central tasks of SpiNNaker will be in analyzing and understanding how brain processing works. It has already been used, on its way to 1 million cores, to run an 80,000-neuron model of a segment of the brain's cortex. It has also simulated the brain's Basal Ganglia region—an area affected by Parkinson's disease. Thus, it could potentially be used in pharmaceutical science. Other similar unprecedentedly large-scale simulations are planned by scientists.

Working on SpiNNaker will also hopefully aid in the realization of mould-breaking new computer architectures with an expected result being a significant leap in energy efficiency in supercomputer systems (Fig. 9.37).[75]

5.26.   Example 103. BrainGate

The technology comes from a project that a few years back began working on Brain-Gate: A way to use the mind to control prosthetic devices using brain power. Here's a sample of the solution in action with a paralyzed patient playing Pong by imagining that they're moving a mouse.

BrainGate requires a small chip implanted into the patient's brain, where it monitors the signals from roughly 100 neurons.

Scientists have narrowed down the section of the brain that controls motor skills, so focusing on that area provides the opportunity to translate those electrical impulses into actions.

---

[75] https://hexus.net/tech/news/systems/123908-worlds-largest-supercomputer-brain-powered-manchester/.

Fig. 9.37 World's largest supercomputer "brain" powered up

Initially, the process to enter text in a Google search box was slow for Patient T6 because the interface was designed to show the entire alphabet.

That required her to look at each letter, similar to how one would type.

The project team decided an interface upgrade could speed up the process so it designed a Bluetooth interface with a Google Nexus 9 tablet.

Essentially, the patient could then simply think about what letter she was tapping on the display much like using a mouse, making the input method far more effective.

Since the interface is now more easily controlled, the patient can "tap" anywhere on the tablet display, including on the hyperlinks of a webpage.

That's a big step forward but the BrainGate team is far from done. Their next project is to add support for click-and-drag as well as multi-touch navigation with the idea being that the mind becomes a more robust computing interface (Fig. 9.38).[76]

5.27.   Example 104. 3D printing of stem cell cartilage

Melbourne surgeons use 3D printer pen filled with stem cells to draw knee cartilage.

MELBOURNE surgeons have used a 3D printer pen filled with stem cell ink to "draw" new cartilage into damaged knees, opening up a world of possibilities for human body part replacements.

The breakthrough Biopen paves the way for the Melbourne-led team eventually to repair damage to bones, muscles, and tendons, and even to tissue in organs such as the heart, the liver, and the lungs.

It has been used in six sheep to repair knee injuries similar to those commonly suffered by Australian rules footballers.

St. Vincent's Hospital orthopedic surgeon Professor Peter Choong said the technique could be adapted to treat a range of conditions in humans.

"The healing was exceptional," Professor Choong said.

"Although we have used this primarily for cartilage, we can already see how this can be used in a variety of other clinical situations."

---

[76] https://www.zdnet.com/article/braingate-als-patient-performs-google-searches-on-nexus-9-tablet-using-her-mind/.

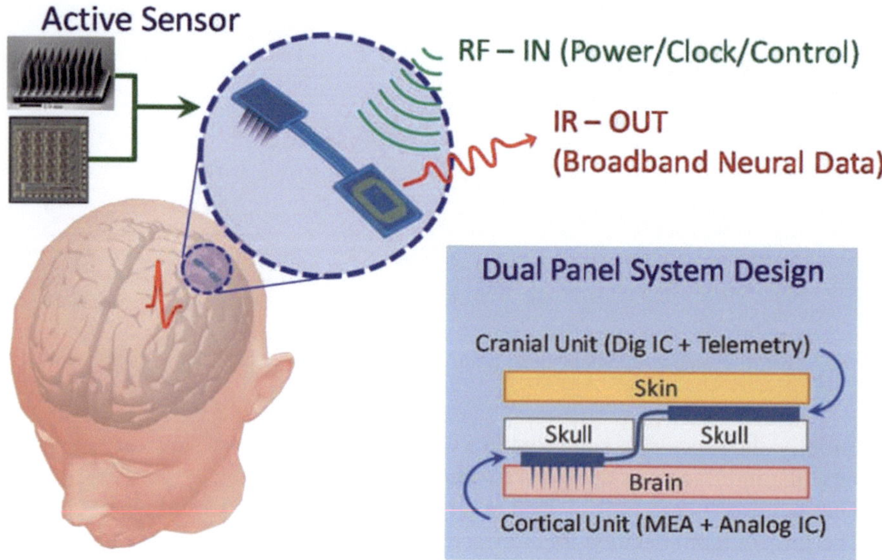

Fig. 9.38 BrainGate[77]

The Biopen was developed by the Aikenhead Centre for Medical Discovery, based at St. Vincent's Hospital, the University of Melbourne, and the University of Wollongong.[78]

A handheld "biopen" capable of 3D printing cartilage tissue could for the first time be used during surgery to treat cartilage injuries and osteoarthritis. The extrusion-based device, which prints live stem cells embedded in a hydrogel material, produces constructs that look and behave just like natural articular tissue (Fig. 9.39).

5.28.   Example 105. See-through walls

Specialists were able to follow the movements of a person behind the wall using only a smartphone.

To spy on a man in his own home, just one smartphone will be enough. This is the conclusion reached by researchers at the University of California, Santa Barbara.

Wi-Fi networks fill the world with radio signals. At work, at home, and especially on the streets, a person is constantly in the field of radio signals in the range from 2.4 to 5 GHz. During movement, it violates and distorts this field, reflecting and refracting radio signals. According to researchers, by analyzing changes in the electromagnetic field, you can determine the location, actions, and movements of a person.

Previously, several research groups have already developed visualization systems that allow "see-through walls" using Wi-Fi, but they had their drawbacks. For

---

[77] https://singularityhub.com/2009/06/17/braingate2-your-mind-just-went-wireless/wireless-bra ingate-diagram/.

[78] https://medicine.unimelb.edu.au/school-structure/surgery/news-and-events/melbourne-sur geons-use-3d-printer-pen-filled-with-stem-cells-to-draw-knee-cartilage.

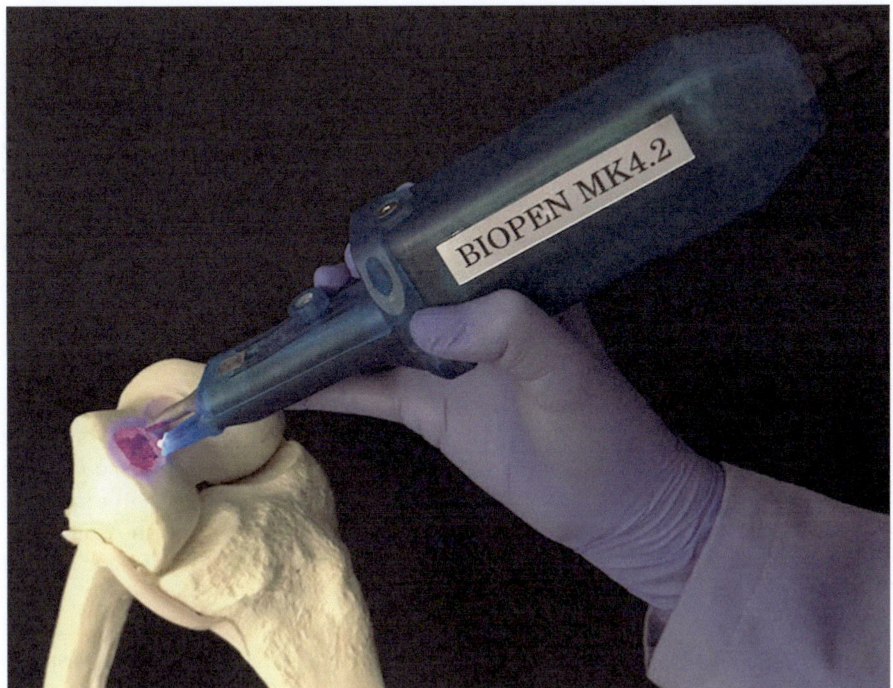

Fig. 9.39 3D printing of stem cell cartilage[79]

example, the attacker had to know the location of the wireless signal transmitter and be authorized on the wireless network.

Researchers at the University of California at Santa Barbara presented a new system that uses surrounding Wi-Fi signals and a regular smartphone for surveillance. "Using a smartphone, attackers can localize and track a person's location in the house or at work through walls using the reflection of surrounding Wi-Fi signals," the researchers explained.

There are no doors or walls for radio signals. They mix, reflecting from different surfaces, and from the point of view of Wi-Fi, the picture of the world looks very blurry. People also reflect signals, however, unlike static objects, they do not look blurry. Therefore, if you look at the world through the eyes of wireless signals, you can easily see the person behind the wall and determine its movements.

To spy on a man behind a wall, the researchers first determined the location of the Wi-Fi transmitter by measuring the signal strength around the house being attacked using a smartphone. If nothing moves inside the house, the signal remains unchanged. However, the slightest movement changes the signal, and each movement does it in its own way.

---

[79] https://physicsworld.com/a/handheld-biopen-prints-human-cartilage/.

The researchers tested their method using Nexus 5 and Nexus 6 smartphones in 11 different offices and apartments. In some of them, there were several Wi-Fi signal transmitters, which increased the accuracy of observation. If there are more than two signal sources in a normal room, the system developed by the researchers detected the presence and movement of people with an accuracy of 99% (Fig. 9.40).[80]

5.29.   Example 106. A breakthrough for soft robots to advance artificial muscles

Scientists have developed a tiny pump that could play a big role in the development of autonomous soft robots, lightweight exoskeletons, and smart clothing. Flexible, silent, and weighing only one gram, it is poised to replace the rigid, noisy, and bulky pumps currently used.

Soft robots have a distinct advantage over their rigid forebears: they can adapt to complex environments, handle fragile objects, and interact safely with humans. Made from silicone, rubber, or other stretchable polymers, they are ideal for use in rehabilitation exoskeletons and robotic clothing. Soft bio-inspired robots could one day be deployed to explore remote or dangerous environments. Most soft robots are actuated by rigid, noisy pumps that push fluids into the machines' moving parts. Because they are connected to these bulky pumps by tubes, these robots have limited autonomy and are cumbersome to wear at best.

Researchers in EPFL's Soft Transducers Laboratory (LMTS) and Laboratory of Intelligent Systems (LIS), in collaboration with researchers at the Shibaura Institute of Technology in Tokyo, Japan, have developed the first entirely soft pump—even the electrodes are flexible. Weighing just one gram, the pump is completely silent and consumes very little power, which it gets from a 2 cm by 2 cm circuit that includes a rechargeable battery. "If we want to actuate larger robots, we connect several pumps together," says Herbert Shea, the director of the LMTS at the School of Engineering.

This innovative pump could rid soft robots of their tethers. "We consider this a paradigm shift in the field of soft robotics," adds Shea. Soft pumps can also be used to circulate liquids in thin flexible tubes embedded in smart clothing, leading to garments that can actively cool or heat different regions of the body. That would meet the needs of surgeons, athletes, and pilots, for example (Fig. 9.41).[81]

5.30.   Example 107. Sensors in the shoe

Engineers at the University of Delaware are developing next-generation smart textiles by creating flexible carbon nanotube composite coatings on a wide range of fibers, including cotton, nylon, and wool. Fabric coated with this sensing technology could be used in future "smart garments" where the sensors are slipped into the soles of shoes or stitched into clothing for detecting human motion.

Carbon nanotubes give this light, flexible, breathable fabric coating impressive sensing capability. When the material is squeezed, large electrical changes in the fabric are easily measured. "As a sensor, it's very sensitive to forces ranging from

---

[80] https://hackernews.blog/wi-fi-signals-allow-you-to-see-through-walls/.

[81] A breakthrough for soft robots to advance artificial muscles https://medicalview.org/a-breakthrough-for-soft-robots-to-advance-artificial-muscles/.

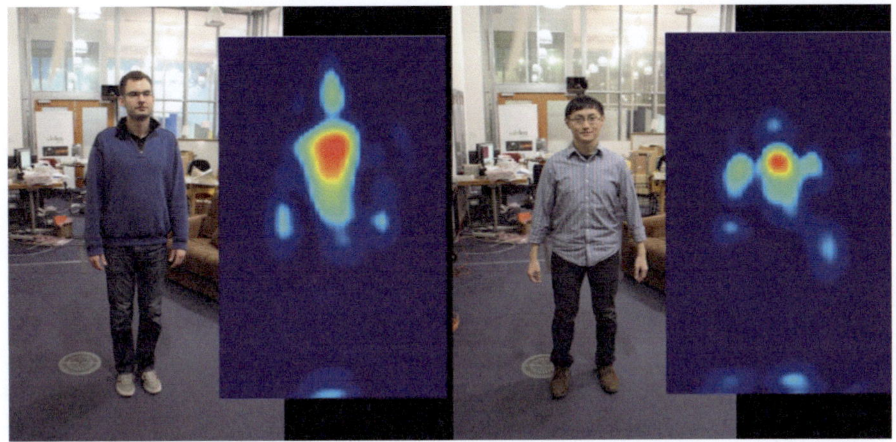

Fig. 9.40 See-through walls[82]

Fig. 9.41 Soft robots to advance artificial muscles[83]

touch to tons," said Erik Thostenson, an associate professor in the Departments of Mechanical Engineering and Materials Science and Engineering (Fig. 9.42).[84]

5.31. Example 108. Wearable sensors detect what's in your sweat

A team of scientists at the University of California, Berkeley, is developing wearable skin sensors that can detect what's in your sweat. Monitoring perspiration could bypass the need for more invasive procedures like blood draws, and provide real-time updates on health problems such as dehydration or fatigue.

---

[82] https://www.techspot.com/news/62606-mit-researchers-develop-technology-can-see-through-walls.html.

[83] *(c) Vito Cacucciolo / 2019 EPFL 2019 EPFL.*

[84] https://medicalview.org/sensors-in-the-shoe-help-with-postural-damage/.

Fig. 9.42 Sensors in the shoe

The team describes a new sensor design that can be rapidly manufactured using a "roll-to-roll" processing technique that essentially prints the sensors onto a sheet of plastic like words on a newspaper.

The new sensors contain a spiraling microscopic tube, or microfluidic, that wicks sweat from the skin. By tracking how fast the sweat moves through the microfluidic, the sensors can report how much a person is sweating, or their sweat rate.

The microfluidics are also outfitted with chemical sensors that can detect concentrations of electrolytes like potassium and sodium, and metabolites like glucose.

They also used the sensors to compare sweat glucose levels and blood glucose levels in healthy and diabetic patients, finding that a single sweat glucose measurement cannot necessarily indicate a person's blood glucose level. "There's been a lot of hope that non-invasive sweat tests could replace blood-based measurements for diagnosing and monitoring diabetes, but we've shown that there isn't a simple, universal correlation between sweat and blood glucose levels," said Mallika Bariya, a graduate student in materials science and engineering at UC Berkeley and the other lead author on the paper. "This is important for the community to know, so that going forward we focus on investigating individualized or multi-parameter correlations (Fig. 9.43)."[85]

5.32.   Example 109. Wireless sensors stick to the skin to track our health

We tend to take our skin's protective function for granted, ignoring its other roles in signaling subtleties like a fluttering heart or a flush of embarrassment. Now, Stanford engineers have developed a way to detect physiological signals emanating from the skin with sensors that stick like band-aids and beam wireless readings to a receiver clipped onto clothing.

To demonstrate this wearable technology, the researchers stuck sensors to the wrist and abdomen of one test subject to monitor the person's pulse and respiration by detecting how their skin stretched and contracted with each heartbeat or breath. Likewise, stickers on the person's elbows and knees tracked arm and leg motions by

---

[85] https://medicalview.org/wearable-sensors-detect-whats-in-your-sweat/.

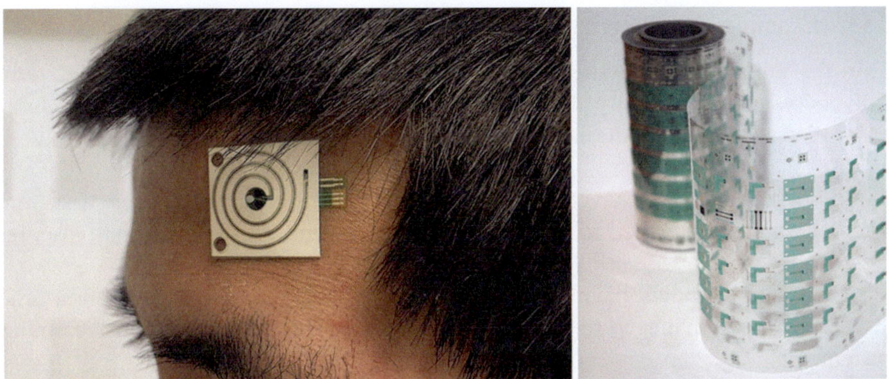

Fig. 9.43 Sensors detect what's in sweat

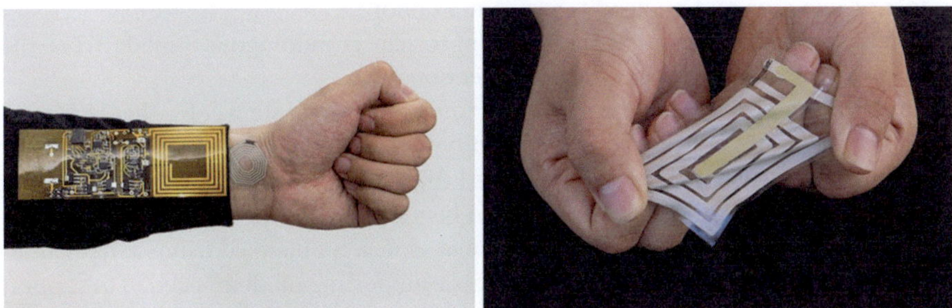

Fig. 9.44 Sensors to track health

gauging the minute tightening or relaxation of the skin each time the corresponding muscle flexed.

Zhenan Bao, the Chemical Engineering Professor whose lab developed the system, thinks this wearable technology, which they call BodyNet, will first be used in medical settings such as monitoring patients with sleep disorders or heart conditions. Her lab is already trying to develop new stickers to sense sweat and other secretions to track variables such as body temperature and stress. Her ultimate goal is to create an array of wireless sensors that stick to the skin and work in conjunction with smart clothing to more accurately track a wider variety of health indicators than the smart phones or watches consumers use today. "We think one day it will be possible to create a full-body skin-sensor array to collect physiological data without interfering with a person's normal behavior," said Bao, who is also the K. K. Lee Professor in the School of Engineering (Fig. 9.44).[86]

---

[86] https://medicalview.org/wireless-sensors-stick-to-the-skin-to-track-our-health/.

5.33.   Example 110. 3D-printed heart with blood vessels

from the laboratory of tissue engineering and regenerative medicine at Tel Aviv University, led by university professor Tal Dvir, announced they had managed a first-of-its-kind achievement: the creation of a miniature but working heart, complete with blood vessels.

While heart models have been 3D printed in the past, this is the first-time researchers have managed to create a proof-of-concept with a functional vascular system.

While heart models have been 3D printed in the past, this is the first-time researchers have managed to create a proof-of-concept with a functional vascular system.

Though still a long way off from being a viable technology, the achievement marks another step towards a future where organs for transplant can be created in a lab, customized for each patient according to their immunological properties, with no fear of rejection and no need to wait for a donor.

This approach ensures that the resulting organ is autologous—made from the same cells—negating the risk of organ rejection.

The harvested cells were reprogrammed into pluripotent stem cells, that is, immature cells that can mature into various cell types, and combined with the a-cellular material into bio-ink that could be used to 3D print tissue. CT scans of a human heart were used to plan the structure of the heart and its larger blood vessels, while smaller blood vessels that cannot be seen on a scan were planned using a mathematical model. The blueprint was then printed in a supporting medium needed to support larger printed structures, and the cells were left to mature.

After a period of maturation, the result was a miniature heart, 20 mm in height and 14 mm in diameter, with the major blood vessels and cells that can contract though not pump.

The ability to print a working blood vessel network is still limited, the researchers noted in the article, and better techniques to image the entire blood vessels of the heart are needed to achieve the next step toward an actual functioning organ that can be transplanted. Current 3D printing technology must also be improved to ensure small blood vessels can be printed accurately.

The researchers are hoping to start testing the hearts in animal models within the year (Fig. 9.45).[87]

5.34.   Example 111. Bioprinting the Human Heart

A team of researchers from Carnegie Mellon University has published a paper in Science that details a new technique allowing anyone to 3D bioprint tissue scaffolds out of collagen, the major structural protein in the human body. This first-of-its-kind

---

[87] Reuven Edri, Idan Gal, Nadav Noor, Tom Harel, Sharon Fleischer, Nofar Adadi, Ori Green, Doron Shabat, Lior Heller, Assaf Shapira, Irit Gat-Viks, Dan Peer, Tal Dvir. Personalized Hydrogels for Engineering Diverse Fully Autologous Tissue Implants. *Advanced Materials,* Mater.2019, 31, 1803895.

https://onlinelibrary.wiley.com/doi/full/10.1002/adma.201803895.

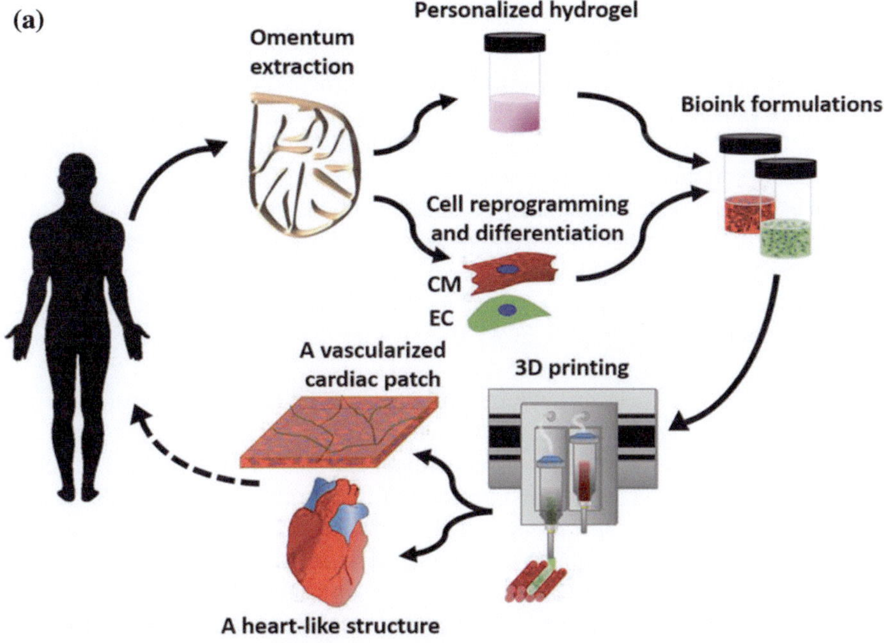

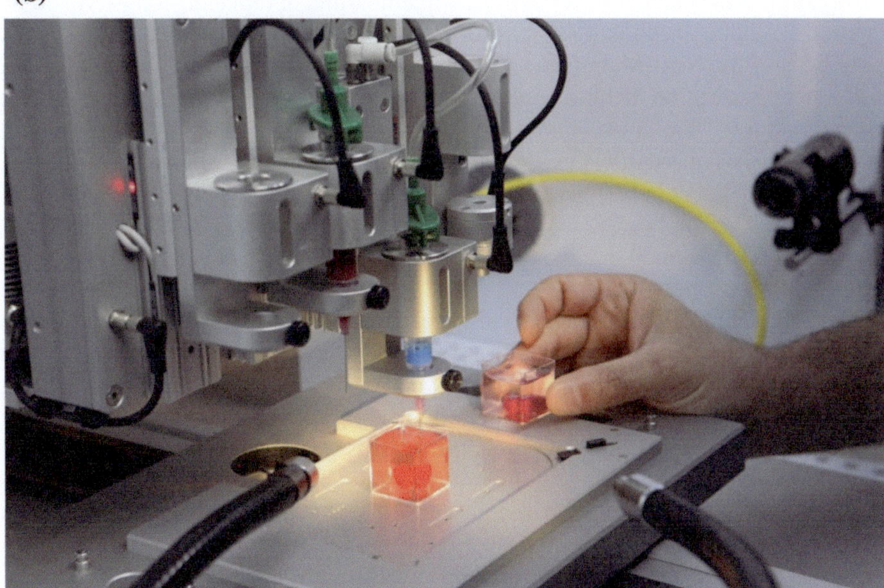

Fig. 9.45 a 3D-printed heart with blood vessels, **b** 3D-printed heart

method brings the field of tissue engineering one step closer to being able to 3D print a full-sized, adult human heart.

The technique, known as Freeform Reversible Embedding of Suspended Hydrogels (FRESH), has allowed the researchers to overcome many challenges associated with existing 3D bioprinting methods, and to achieve unprecedented resolution and fidelity using soft and living materials.

Each of the organs in the human body, such as the heart, is built from specialized cells that are held together by a biological scaffold called the extracellular matrix (ECM). This network of ECM proteins provides the structure and biochemical signals that cells need to carry out their normal function. However, until now it has not been possible to rebuild this complex ECM architecture using traditional biofabrication methods.

"What we've shown is that we can print pieces of the heart out of cells and collagen into parts that truly function, like a heart valve or a small beating ventricle," said Adam Feinberg, a professor of biomedical engineering (BME) and materials science and engineering, whose lab performed this work. "By using MRI data of a human heart, we were able to accurately reproduce patient-specific anatomical structure and 3D bioprint collagen and human heart cells."

Over 4,000 patients in the United States are waiting for a heart transplant, while millions of others worldwide need hearts but are ineligible for the waitlist. The need for replacement organs is immense, and new approaches are needed to engineer artificial organs that are capable of repairing, supplementing, or replacing long-term organ function. Feinberg, who is a member of Carnegie Mellon's Bioengineered Organs Initiative and the Next Manufacturing Center, is working to solve these challenges with a new generation of bioengineered organs that more closely replicate natural organ structures. "Collagen is an extremely desirable biomaterial to 3D print with because it makes up literally every single tissue in your body," said Andrew Hudson, a BME Ph.D. student in Feinberg's lab and co-first author of the paper. "What makes it so hard to 3D print, however, is that it starts out as a fluid—so if you try to print this in air it just forms a puddle on your build platform. So, we've developed a technique that prevents it from deforming."

The FRESH 3D bioprinting method developed in Feinberg's lab allows collagen to be deposited layer-by-layer within a support bath of gel, giving the collagen a chance to solidify in place before it is removed from the support bath. With FRESH, the support gel can be easily melted away by heating the gel from room temperature to body temperature after the print is complete. This way, the researchers can remove the support gel without damaging the printed structure made of collagen or cells.

This method is truly exciting for the field of 3D bioprinting because it allows collagen scaffolds to be printed at the large scale of human organs. It is not limited to collagen, as a wide range of other soft gels including fibrin, alginate, and hyaluronic acid can be 3D bioprinted using the FRESH technique, providing a robust and adaptable tissue engineering platform. Importantly, the researchers also developed open-source designs so that nearly anyone, from medical labs to high school science classes, can build and have access to low-cost, high-performance 3D bioprinters.

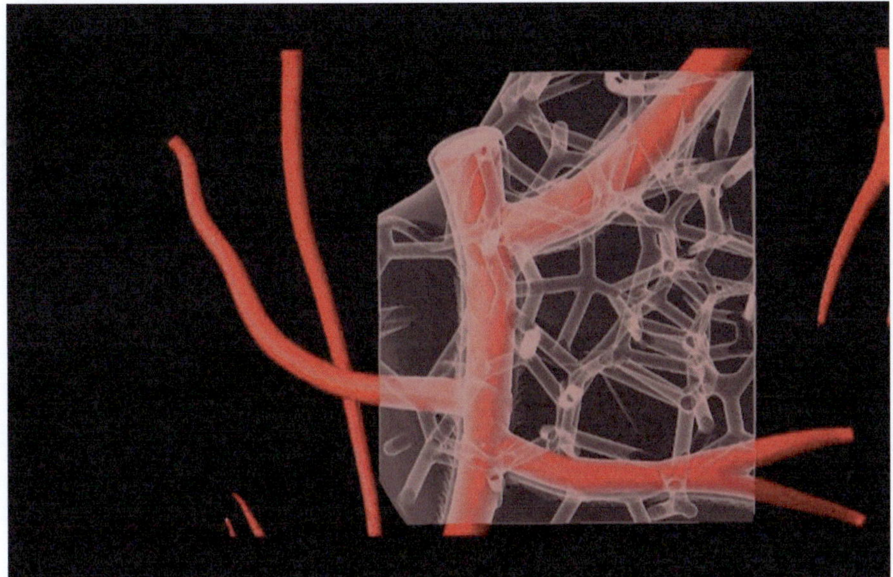

Fig. 9.46 Bioprinting the human heart

Looking forward, FRESH has applications in many aspects of regenerative medicine, from wound repair to organ bioengineering, but it is just one piece of a growing biofabrication field. "Really what we're talking about is the convergence of technologies," Feinberg said. "Not just what my lab does in bioprinting, but also from other labs and small companies in the areas of stem cell science, machine learning, and computer simulation, as well as new 3D bioprinting hardware and software."

"It is important to understand that there are many years of research yet to be done," Feinberg added, "but there should still be excitement that we're making real progress towards engineering functional human tissues and organs, and this paper is one step along that path (Fig. 9.46)."[88]

5.35. Example 112. The organ-on-e-chip

Researchers from Carnegie Mellon University (CMU) and Nanyang Technological University, Singapore (NTU Singapore) have developed an organ-on-an-electronic-chip platform, which uses bioelectrical sensors to measure the electrophysiology of heart cells in three dimensions.

These 3D, self-rolling biosensor arrays coil up over heart cell spheroid tissues to form an "organ-on-e-chip," thus enabling the researchers to study how cells communicate with each other in multicellular systems such as the heart.

The organ-on-e-chip approach will help develop and assess the efficacy of drugs for disease treatment—perhaps even enabling researchers to screen for drugs and

---

[88] https://medicalview.org/one-step-closer-to-bioprinting-the-human-heart/.

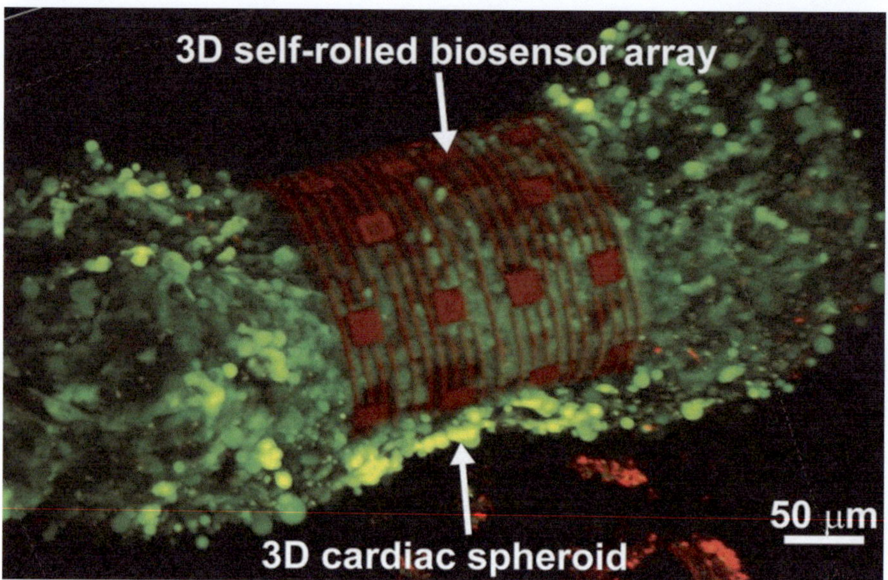

Fig. 9.47   The organ-on-e-chip

toxins directly on a human-like tissue, rather than testing on animal tissue. The platform will be used to shed light on the connection between the heart's electrical signals and disease, such as arrhythmias. The research allows the researchers to investigate processes in cultured cells that currently are not accessible, such as tissue development and cell maturation.

"For decades, electrophysiology was done using cells and cultures on two-dimensional surfaces, such as culture dishes," said Tzahi Cohen-Karni, an Associate Professor of Biomedical Engineering and Materials Science and Engineering. "We are trying to circumvent the challenge of reading the heart's electrical patterns in 3D by developing a way to shrink-wrap sensors around heart cells and extracting electrophysiological information from this tissue (Fig. 9.47)."[89]

5.36.   Example 113. Robotic thread to clear stroke clots

MIT engineers have developed a magnetically steerable, thread-like robot that can actively glide through narrow, winding pathways, such as the labyrinthine vasculature of the brain. In the future, this robotic thread may be paired with existing endovascular technologies, enabling doctors to remotely guide the robot through a patient's brain vessels to quickly treat blockages and lesions, such as those that occur in aneurysms and stroke.

"Stroke is the number five cause of death and a leading cause of disability in the United States. If acute stroke can be treated within the first 90 minutes or so, patients'

---

[89] Self-dolling sensor take heart cell readings in 3D https://medicalview.org/self-rolling-sensors-take-heart-cell-readings-in-3d/.

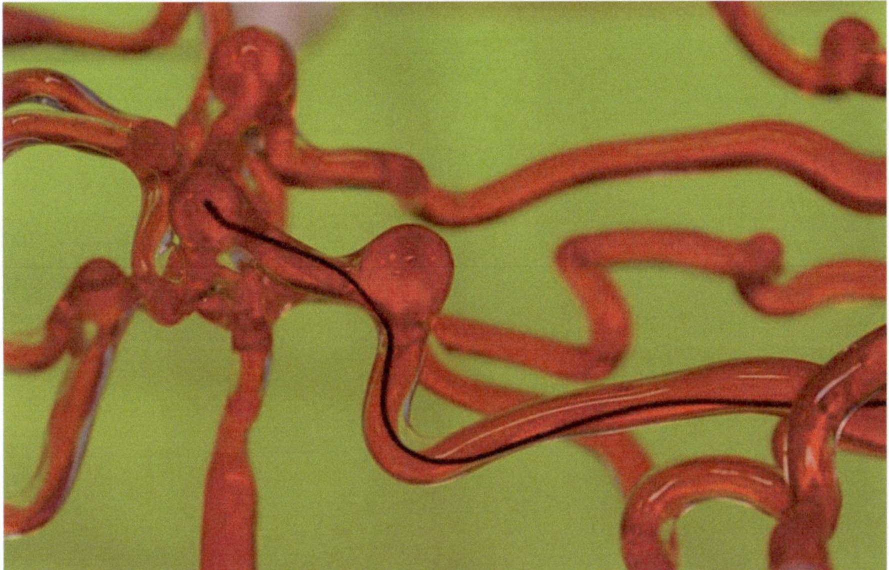

Fig. 9.48 Robotic thread to clear stroke clots

survival rates could increase significantly," says Xuanhe Zhao, Associate Professor of Mechanical Engineering and of Civil and Environmental Engineering at MIT. "If we could design a device to reverse blood vessel blockage within this 'golden hour,' we could potentially avoid permanent brain damage. That's our hope (Fig. 9.48)."[90]

5.37.   Example 114. A wearable device for brain monitoring

Imagine if a coach could know which moments of competition a certain player might peak, or if a truck driver had objective data telling him his body and mind were too tired to continue driving. Traditionally, measuring alertness or mental fatigue requires interrupting a natural moment to intervene in an artificial setting. But Penn neuroscientist Michael Platt and postdoc Arjun Ramakrishnan have created a tool to use outside the lab, a wearable technology that monitors brain activity and sends back data without benching a player or asking a trucker to pull over.

The platform is akin to a Fitbit for the brain, with a set of silicon and silver nanowire sensors embedded into a head covering like a headband, helmet, or cap. The device, a portable electroencephalogram (EEG), is intentionally unobtrusive to allow for extended wear, and, on the backend, powerful algorithms decode the brain signals the sensors collect. Though it's still in the early stages, the technology has potential applications from health care to sports performance and customer engagement (Fig. 9.49).[91]

---

[90] https://medicalview.org/robotic-thread-to-clear-stroke-clots/.

[91] https://medicalview.org/a-wearable-device-for-brain-monitoring/.

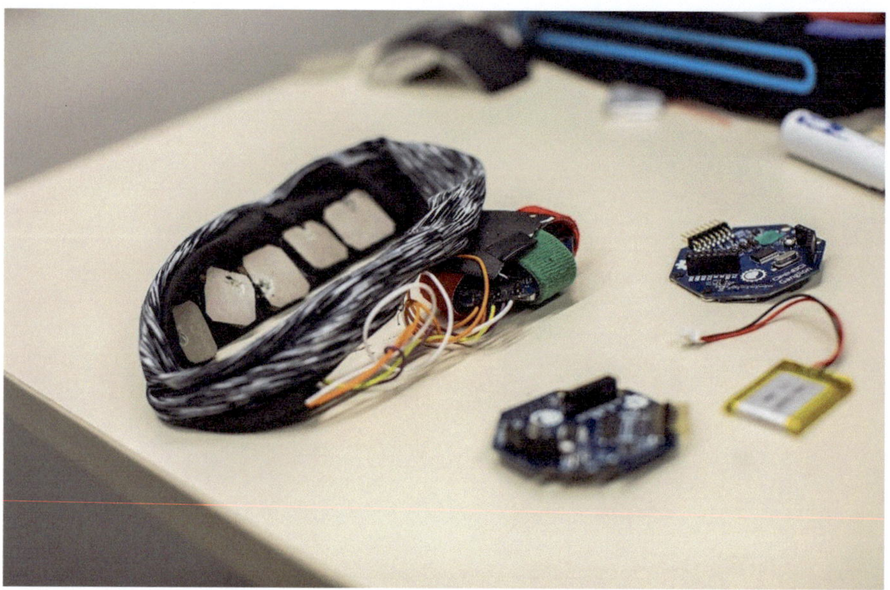

Fig. 9.49 A wearable device for brain monitoring

6. *Solve the problems using Su-Field analysis*

6.1. Problem 15. Golden chain

Condition of the problem

At the jewel manufacturing plant, the links of the gold chain are welded on automatic machines. Such chains are very strong: even the thinnest and the most exquisite chains cannot be torn with hands. Such a chain can become a «strangler» for its owners (for example, in case of robbery or in other cases).

What's to be done?

6.2. Problem 16. **Problem 17. Warming bottle**

Conditions of the problem

Coupling «B», which includes a metal bar «C» is tightly pressed into the element «A» (Fig. 9.50).[92]

How to extract coupling «B», using the simplest tools, for example, hammer?

---

[92] The problem and its analysis was formulated by the author in: A. N. Garina-Domchenko, A. S. Galysheva, V. M. Petrov, B. L.Zlotin, S. S. Litvin, V. F. Kaner. Training and methodological materials on the basics of technical creativity for vocational schools.—L.: Research Institute of Professional Technical Education. 1979.—212 p. (P. 187–188).

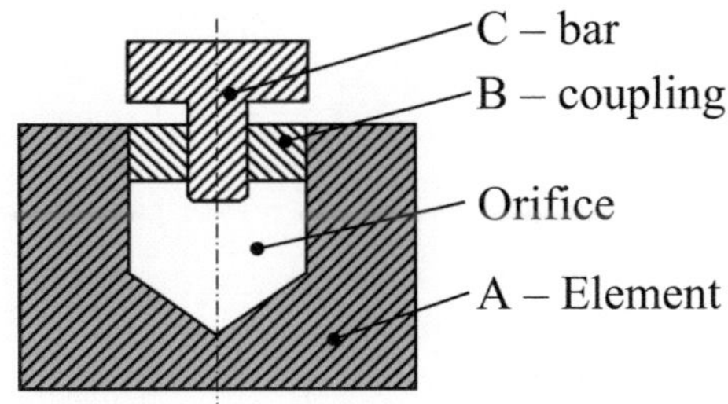

Fig. 9.50 Extraction of pressed-in coupling

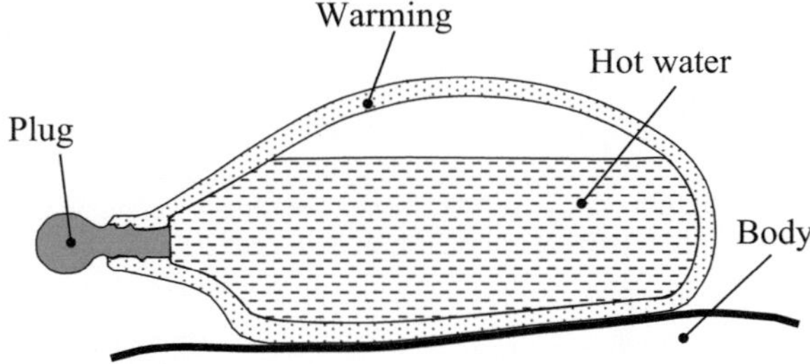

Fig. 9.51 Warming bottle

6.3.   Problem 17. Warming bottle

Conditions of the problem

A warming bottle (Fig. 9.51), into which boiling water is just poured, can burn the patient.

What's to be done?

6.4.   Problem 18. Abrasive processing

Condition of the problem

The apparatus for abrasive processing of component parts of complicated shape consists of two pipes, which are arranged coaxially. Air moves along the internal pipe, while particles of abrasive material move along the external pipe. At the end of the external pipe, a nozzle is located, which forms the jet of the abrasive material (Fig. 9.52). The nozzle quickly gets worn out, and it is necessary to replace it. What's to be done, in order to create a nozzle, which does not wear out?

Usually, they try to manufacture nozzles of more wear-proof materials, however, even they get worn out, while the cost of such materials is much higher.

6.5.   Problem 19. Packing the products

Conditions of the problem

A method is known for packing and conservation of products by submerging them into a polymer melt (Fig. 9.53). It is fairly difficult to take off such a pack of products,

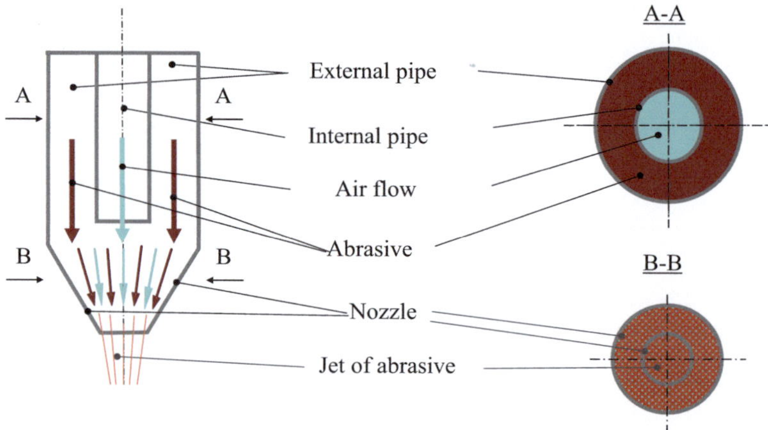

Fig. 9.52 Apparatus for abrasive processing of components. Where A-A—a cross-section of coaxial pipes, B-B—a cross-section of the nozzle

Fig. 9.53 Packing products with a complicated relief

the surface of which has a complicated relief. It is necessary to cut it, which can lead to damage to the product's surface.

What's to be done?

6.6.   Problem 20. Supports of the bridges

Conditions of the problem

In winter time the supports of the bridges are covered with ice and are gradually destroyed. Water gets into the cracks formed in the supports. As a result of freezing ice is expanded and splits pieces off.

It is necessary to eliminate this disadvantage in the simplest and cheapest way.

Appendix A
Analysis of Examples and Problems

In this section, we shall show you our variant of analysis of some examples and problems.

Example 41. Turbine of a jet engine

The turbines of jet engines operate at high-temperature values. In order to preserve the strength characteristics of turbine blades, it is necessary to add alloy additives to the source material, for example, cobalt, which significantly increases the cost of the turbine, however, confers resistance to high temperature to it. Pratt & Whitney developed a technology for blade manufacturing, which enables to reduce the content of cobalt in them by 30%. To achieve this, they drill very small holes in the blades with alloy. The air passing through the holes, cools the blades, and besides, aerodynamic resistance is reduced. Thus, the turbines can be manufactured of a material, which is less heat-resistant.[1]

Analysis of example

The solution could be presented as a Su-field formula (A.1).

$$S_1 \rightleftharpoons S_2 \implies S_1 \longrightarrow S_2 \qquad S_3 = S_2' \tag{A.1}$$

where

S_1 air
F_1 temperature field (high temperature)
S_2 blade
S_2' blade with holes

---

[1] Inventor and rationalizer, No. 12, 1985, Micro Information 1203.—P. 30.

V. Petrov, *Structural System Analysis*, https://doi.org/10.1007/978-3-031-55825-2

Condition—prototype

Temperature F_1 heats air S_1, which positively acts upon the blade (generates force, which rotates the turbine) and negatively acts upon the blade (destroys it).

A Su-Field is given with a useful and harmful bond between S_1 and S_2.

Solution

In order to eliminate harmful bond in keeping with model (A.1) between substances S_1 and S_2, it is necessary to place the substance S_3, which constitutes either these very substances or their modifications: S_1' or S_2'. The solution is based on selecting the modification of blade S_2'—blade with orifices.

Example 42. Elimination of cavitation

Cavitation causes erosion (destruction) of the material of the devices, where it occurs. Attempts are made to eliminate cavitation. In this case, it is very important that the cavitation should be eliminated uniformly. It is suggested to act upon cavitation bubbles with ultrasound oscillations in the range of frequencies from 1 to 50 kHz.[2]

Analysis of example

Solution could be presented in the form of a Su-field formula (A.2).

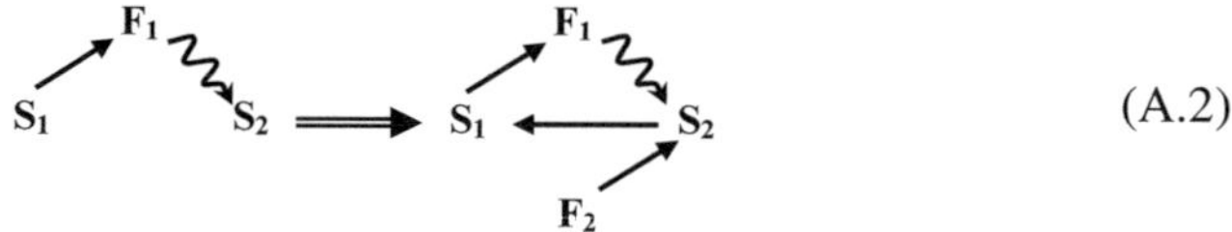

$$(A.2)$$

where

S_1 cavitation bubble
F_1 field of micro-explosion, destroying the material of the device
B_2 material of the device
F_2 ultrasound field

Condition—prototype

Cavitation bubble S_1, after collapsing on the material of the device, generates a micro-explosion F_1, causing erosion.

A Su-field is given with a harmful bond between F_1 and S_2.

Solution

In order to eliminate harmful bonds in keeping with the model (A.2) the second field F_2 is introduced, which destroys the harmful action of the field F_1. It is necessary to select the appropriate kind of field F_2, which could counteract the field F_1—micro-explosions, i.e., a field, which would destroy cavitation bubbles S_1.

---

[2] Author's certificate 954 597.

Example 43. Power measuring

Calorimetric method for measuring power. In order to measure power absorbed by the load in the super-high-frequency (SHF) range, the amount of heat is determined, which is given by the load to the operating fluid (water), and the working medium is used as a load. With the aid of the measuring unit, the temperature of the working medium is registered and its value is used for determining the value of power.[3]

Analysis of the example

The object of analysis is the measuring system.

A solution according to formula (A.3) is described

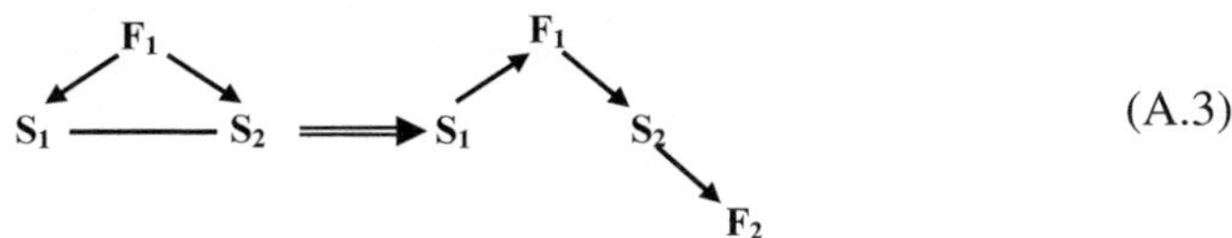

$$(A.3)$$

where

S_1 super-high-frequency (SHF) generator
S_2 load; in the solution the load is water
F_1 electromagnetic field (super-high-frequency field), power of which it is necessary to measure
F_2 temperature

The device S_1 emits an ultra-high-frequency (UHF) field F_1, consumed power of which it is necessary to measure. They introduce water S_2, which is heated by the SHF field, being a load for the device. The power value is determined through temperature.

The solution could also be represented by another formula (A.4).

$$(A.4)$$

where

S_1 SHF Generator
S_2 load
F_1 electromagnetic field (UHF-field)
S_3 Operating medium (water)
F_2 Thermal field
S_4 Sensor of water temperature

---

[3] Yelizarov A. S. Electric radio measurements.—Minsk: Vysheishaya shkola, 1986, 320 p.

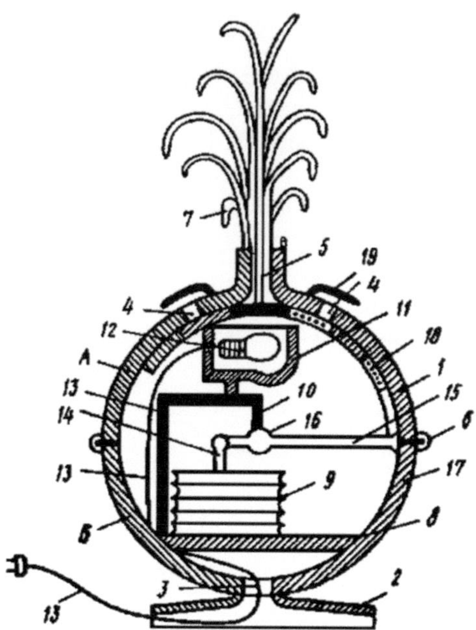

Fig. A.1 Decorative lighting fixture. Author's certificate 779,726. 1—spherical body made of heat resistant plastic, consisting of upper hemisphere A and lower hemisphere B; 2—support; 3—orifice in the lower part of the body; 4—ventilation orifice; 5—throat; 6—annular bracket, connecting halves A and B; 7—beam mode waveguides, built into the throat 5; 8—the substrate; 9—hermetically sealed corrugated vacuum chamber; 10—holder; 11—reflector; 12—source of light; 13—electric power cable; 14—guide path; 15—lever; 16—hinge; 17—fastening of the lever; 18—color filter (light conducting plate, subdivided into individual color sectors); 19—light-reflecting protective screens

Example 44. Decorative lighting fixture

Decorative lighting fixtures are known, which use optical fiber. Such a lighting fixture consists of a lamp, reflector, temperature philter and light philter, connecting head, and optic fiber cable. This lighting fixture had only one philter, which was rigidly fastened.

Compile a Su-Field formula.

A decorative lighting fixture was invented, which changes its color with the variation of atmospheric pressure (Fig. A.1). In this invention, the color filters are fastened on a corrugated vacuum chamber, which changes its volume depending on atmospheric pressure and moves color filters of different colors.[4]

Analysis of example

The solution is described according to model (A.5).

---

[4] Author's certificate 779,726.

$$S_1 \longrightarrow S_2 \implies S_1 \longrightarrow S_2 \longleftarrow S_3 \qquad (A.5)$$

where

S_1 lamp
S_2 light filter (light filter)
F_1 light—optical field
S_3 camera
F_2 atmospheric pressure

Problem 15. Golden chain

Condition of the problem

At the jewel manufacturing plant, the links of the gold chain are welded on automatic machines. Such chains are very strong: even the thinnest and the most exquisite chains cannot be torn with hands. Such a chain can become a «strangler» for its owners (for example, in case of robbery or in other cases).

What's to be done?
What's to be done?

Analysis of the problem

*Used tool—**idealization***: transfer of harmful action to the area, which is prepared in advance by creating *easily damaged sections.*

It is necessary to create *a weak link.*

Solution

The jewelers waken one of the links of a golden chain, soldering it incompletely. In case of necessity, the chain is torn particularly in this place.

The most widely spread service of jewelry workshops is the repair of torn chains.[5]

Problem 16. Extraction of pressed-in coupling

Conditions of the problem

Coupling «B», which includes a metal bar «C» is tightly pressed into the element «A» (Fig. A.2).[6]

How to extract coupling «B», using the simplest tools, for example, hammer?

---

[5] Davydova Ye. My gold//Magazine «Profile». September 25.—2005.—№ 35.—P. 129.

[6] The problem and its analysis was formulated by the author in: A. N. Garina-Domchenko, A. S. Galysheva, V. M. Petrov, B. L. Zlotin, S. S. Litvin, V. F. Kaner. Training and methodological materials on the basics of technical creativity for vocational schools.—L.: Research Institute of Professional Techncial Education VNII. 1979.—212 p. (P. 187–188).

Fig. A.2 Extraction of
pressed-in coupling

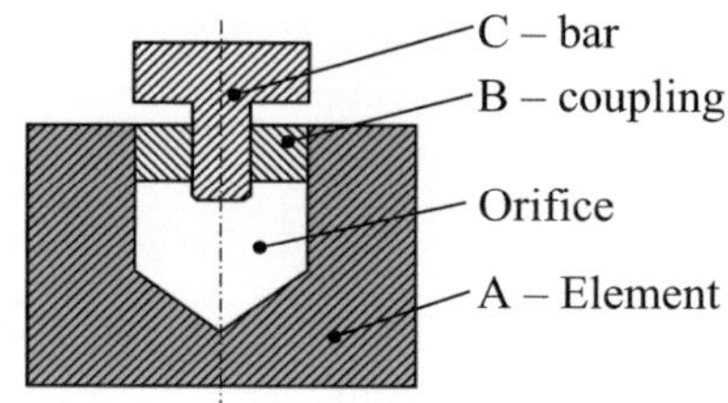

Analysis of the problem

Condition—prototype

Coupling S_2, which contains a metal bar S_3 is tightly pressed into a component S_1 (Fig. A.2).

Solution

Su-field pattern of the problem could be presented by the formula (A.6).

$$
\begin{array}{c}
F_1 \\
S_1 \longrightarrow S_2, S_3
\end{array}
\tag{A.6}
$$

where

S_1 element
S_2 coupling
S_3 bar
F_1 mechanical field (friction), holding the coupling in the element

Poorly controllable Su-Field is given.
Solution could be presented in the form of a Su-Field formula (A.7).

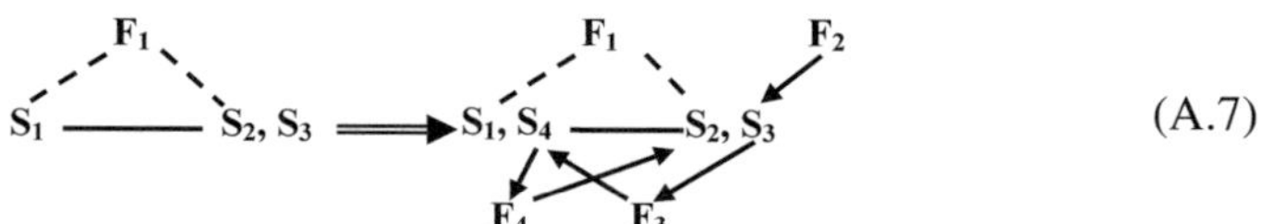

$$
\tag{A.7}
$$

where

S_1 element
S_2 coupling
S_3 bar
F_1 mechanical field (friction), holding the coupling inside the element
S_4 machine oil
F_2 mechanical field (mechanical blow on the bar S_3)
F_3 mechanical field (pressure in machine oil S_4)
F_4 mechanical field (hydraulic shock on coupling S_2)

Fig. A.3 Extraction of pressed-in coupling

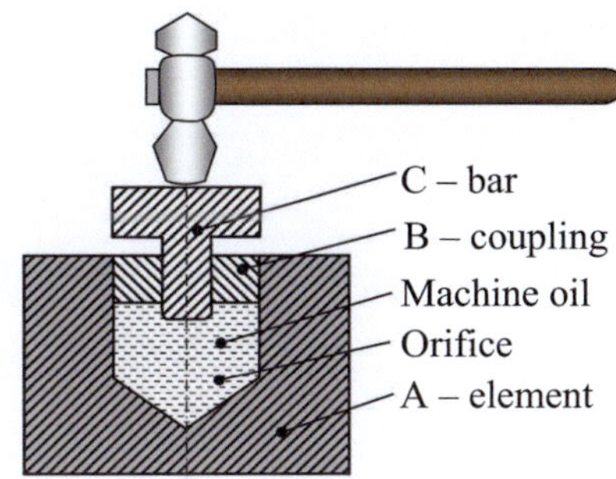

Fig. A.4 Warming bottle

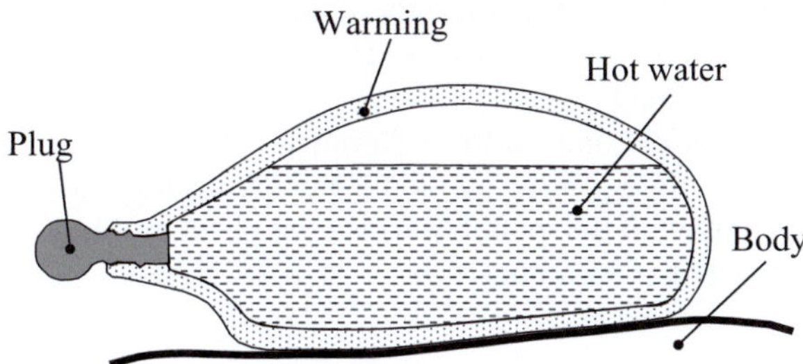

The internal space of the orifice in the element C_1 is completely poured with machine oil S_4. A hammer blow F_2 is exerted on the bar S_3 (Fig. A.3). Pressure F_3 is generated in oil S_4 and hydraulic shock F_4 takes place. The generated oil pressure pulls the coupling from the orifice.

Problem 17. Warming bottle

Conditions of the problem

A warming bottle (Fig. A.4), into which boiling water is just poured, can burn the patient.

What's to be done?

Analysis of the problem

Let us present the problem in a Su-Field form (A.8).

$$(S_1, S_2) \overset{F_1}{\leadsto} S_3 \tag{A.8}$$

where

S_1 ater
S_2 hot water bottle
S_3 body of the patient
F_1 temperature

Fig. A.5 Warming bottle

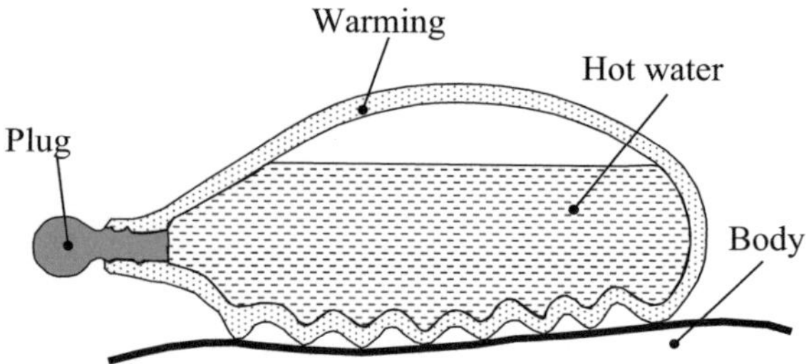

Water S_1 was heated with the aid of a temperature field F_1. Hot water heats the warming bottle S_2, while the latte heats the body of patient S_3 (direct arrow) and can burn him (wavy arrow). This is an internal complex Su-Field with a useful and harmful bond.

Harmful bond could be eliminated by the introduction of S_4, which can take the form of S_1, S_2, S_3 or by modifying them so that they would become S_1', S_2', S_3'. In keeping with the model (4.4) for the given problem, the structural solution could be presented as formula (A.9).

$$(S_1, S_2) \xrightarrow{\quad\wedge\wedge\wedge\quad} S_3 \overset{F_1}{\Longrightarrow} (S_1, S_2) \overset{F_1}{-\!\!\!\mid\!\!\!\rightarrow} S_3 \qquad (A.9)$$

$$S_4 = S_1, S_2, S_3$$
$$S_4 = S_1', S_2', S_3'$$

A warming bottle was patented in Germany, on one side of which there are bulges (Fig. A.5). Thus, only some points of the warming bottle come in contact with the body and there is a layer of air between the body and the warming bottle.

Modification of warming bottle S_2' was used in the present invention.

Problem 18. Abrasive processing

Condition of the problem

The apparatus for abrasive processing of component parts of complicated shape consists of two pipes, which are arranged coaxially. Air moves along the internal pipe, while particles of abrasive material move along the external pipe. At the end of the external pipe, a nozzle is located, which forms the jet of the abrasive material (Fig. 9.49). The nozzle quickly gets worn out, and it is necessary to replace it. What's to be done, in order to create a nozzle, which does not wear out? (Fig. A.6).

Usually, they try to manufacture the nozzle of materials with higher resistance to wearing, however, even if they wear out, the cost of such materials is much higher.

Analysis of the problem

Su-field pattern of the problem could be presented as the following formula (A.10).

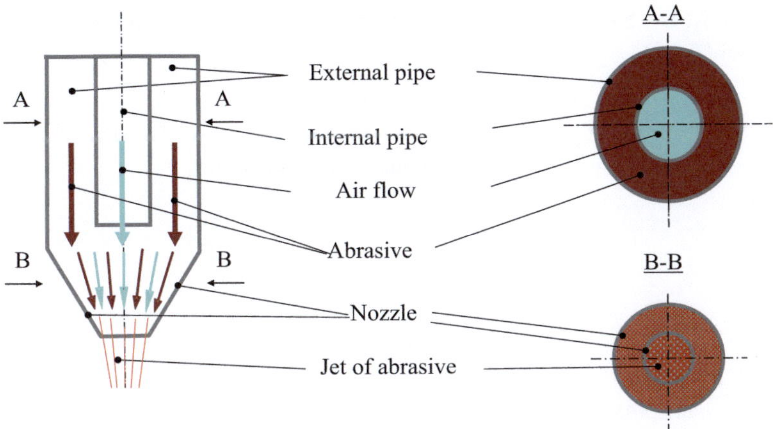

Fig. A.6 Apparatus for abrasive processing of components where A-A—cross-section of coaxial pipes, B-B—cross-section of the nozzle

$$\tag{A.10}$$

where

S_1 abrasive
S_2 nozzle
F_1 air pressure (air flow)

The problem is described by a Su-field with a useful and harmful bond. Useful action (direct line from S_2 to S_1)—formation of abrasive jet. Harmful (wavy line from S_1 to S_2) means the wear-out of the nozzle.

Formula (A.10) could be more exactly presented by formula (A.11).

$$\tag{A.11}$$

where

S_1 abrasive
S_2 nozzle
F_1 pressure of air (air flow)
S_3 air

Possible solution is the use of a tendency to eliminate harmful bond (Fig. 3.12).

One of the solutions in keeping with the model (4.4): between the substances (S_1, S_3) and S_2 the third substance S_4 is introduced, which is either one of the available substances S_1, S_2, S_3 or their modification S_1', S_2', S_3'. This solution is presented by a model (A.12).

$$S_1, S_3 \rightrightarrows S_2 \overset{F_1}{\Longrightarrow} S_1, S_3 \overset{F_1}{\dashv} B_2 \qquad (A.12)$$

$$S_4 = S_1, S_2, S_3$$
$$S_4 = S_1', S_2', S_3'$$

where

S_1	abrasive
S_2	nozzle
F_1	air pressure (airflow)
S_3	air
S_1', S_2', S_3'	modifications S_1, S_2, S_3

Solutions

1.   Nozzle S_2 should hold on its internal surface particles of abrasive S_1
1.1.   Particles of abrasive S_1 are held on the internal surface of the nozzle S_2 with the aid of a **vacuum**.

Solution 1. The nozzle has the form of a meshwork, on which suction (vacuum) is generated. Particles of abrasive are attracted to the meshwork (Fig. A.7).[7] Now the nozzle (meshwork) «are protected» by particles of abrasive. When the particles are worn out, new particles from the flow appear in their place.

Vacuum in this invention is created due to available airflow. For this purpose, channel 8 is provided for (Fig. A.7). The mechanism of the physical phenomenon of **ejection**[8] is shown in Fig. A.8. The flow of gas or fluid passing perpendicularly to the end of the pipe, generates suction (vacuum) in it.

The function of S_4 in the present solution is performed by:

S_2'	modification of the nozzle—meshwork
S_1	abrasive
S_3'	modification of the air—vacuum obtained with the aid of ejection

1.2.   Particles are held due to modification of the nozzle shape.

Solution 2. «Pockets» 10 for abrasive[9] can be made in the nozzle (Fig. A.9a). In this case, the jet of abrasive will rub against particles of stuck abrasive, and stuck

---

[7] Author's certificate 971,639.

[8] **Ejector** (*French*: éjecteur, derived from éjecter—throw away)—a device, in which transfer of kinetic energy from one medium moving at a high speed, to another medium takes place. Transfer of energy takes place in the course of mixing of mediums. **Ejector** is used in jet and vacuum pumps. It is broadly used in chemical and oil processing industries as a mixer.

[9] Author's certificate 1,184,653.

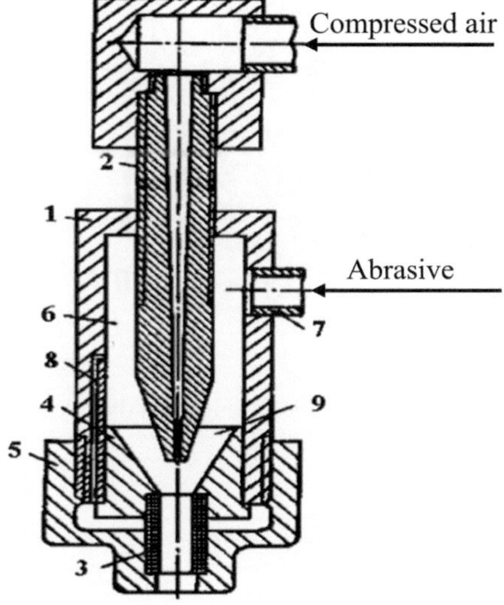

Fig. A.7 Abrasive processing. Author's certificate 971,639. 1—body; 2—air nozzle; 3—insert (embodied of meshwork); 4—coupling; 5—nut; 6—depression chamber; 7—pipeline; 8—channel; 9—mixing chamber

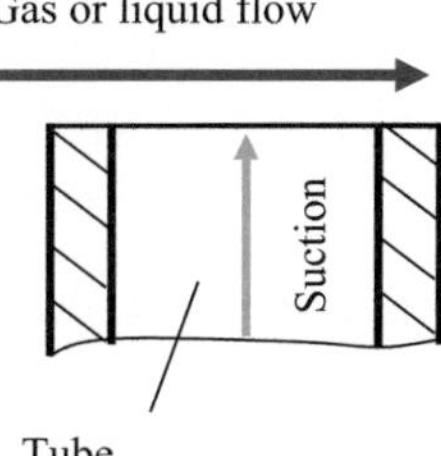

Fig. A.8 Ejector

particles will protect the nozzle from wearing out (Fig. A.9b). The rest is similar to cl. 1.1.

The function of S_4 in this solution is played by

S_2' modification of nozzle—«pockets»
S_1 abrasive

2. Particles of abrasive should not approach the nozzle walls or be pushed away from them.

Solution 3. In the nozzle walls, there are guidelines for compressed air. They are located tangentially with an incline toward the outlet of the nozzle (Fig. A.10). Compressed air is fed through the guidelines. It pushes particles of the abrasive away from the nozzle walls.

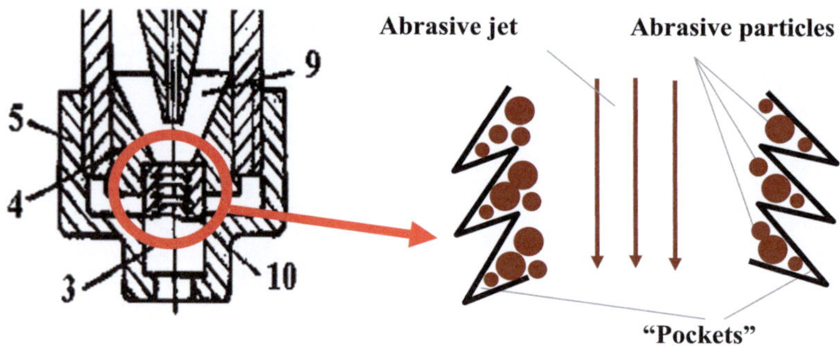

Fig. A.9 Pockets according to Author's certificate 1,184,653

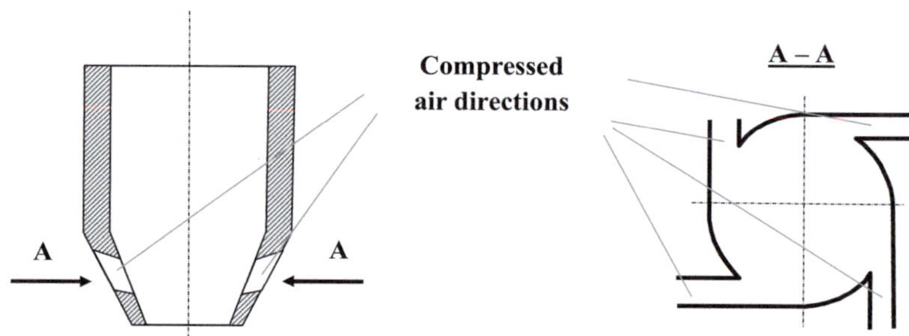

Fig. A.10 Compressed air pushes the particles away

Besides, air jets swirl the flow of abrasive and form the jet. In case with a certain design and air pressure it is possible to reject the main air flow.

The function of S_4 in this solution is played by

S_2' variation of the nozzle—guidelines for compressed air
S_1 abrasive
S_3 air

Solution 4. The simplest way is to change places of air and abrasive (Fig. A.11).

3.   Solutions according to other formulas are also possible, for example (A.13).

$$S_1, S_3 \xrightarrow{\quad\quad} S_2 \Longrightarrow (S_1, S_5)S_3 \xrightarrow[\quad]{F_1} S_2' \searrow F_2 \quad S_4 = S_5 \qquad (A.13)$$

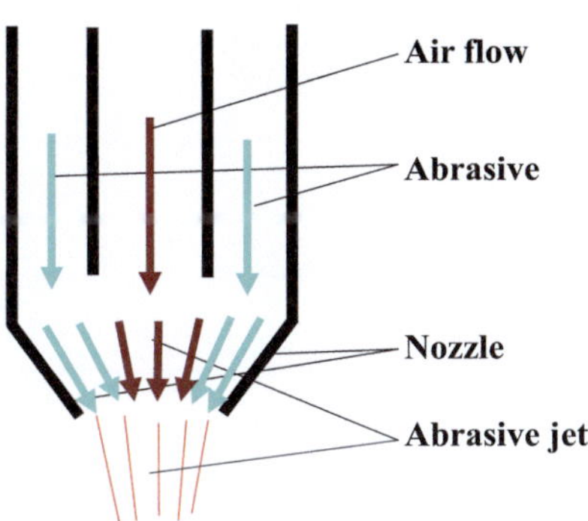

Fig. A.11 Nozzle with abrasive

where

S_1 abrasive
S_2 nozzle
F_1 air pressure (air flow)
S_3 air
S_2' modified nozzle S_2—magnet
S_5 ferromagnetic particles, which stay inside the abrasive
F_2 magnetic field

3.1. Particles of abrasive B_1 are held on the internal surface of the nozzle S_2 with the aid of a magnetic field.

Solution 5. Particles of abrasive S_1 are sintered with ferromagnetic particles of S_5. The nozzle S_2 is modified to the condition of S_2', by embodying it of magnet, which generates a magnetic field, attracting particles of abrasive (S_1, S_5) to the nozzle-magnet S_2'. The rest is similar to cl. 1.1.

The function of S_4 in this solution is performed by

S_2' modification of nozzle—magnet
F_2 magnetic field, generated by magnet S_2'
(S_1, S_5) abrasive, sintered with ferromagnetic particles of S_5

Problem 19. Packing of the products

Conditions of the problem

A method is known for packing and conservation of products by submerging them into a polymer melt. It is fairly difficult to take off such a pack off the products, the surface of which has a complicated relief. It is necessary to cut it, which can lead to damage to the product's surface.

What's to be done? (Fig. A.12).

Fig. A.12 Packing products with a complicated relief

Analysis of the problem

Let us create a Su-Field model of the described system. It could be described by the formula (A.14).

$$S_1 \nwarrow\!\!\!\rightsquigarrow\overset{F}{}\searrow S_2 \qquad\qquad (A.14)$$

Here

S_1 product
S_2 pack
F force, which holds the pack on the product

The problem is described by the Su-Field with useful and harmful bonds. Useful action (direct arrow f F to S_1)—holding the pack on the product. Harmful (wavy line from F to S_1)—inability to take the pack off the product because of strong adhesion.

One of the opportunities for eliminating the described disadvantages is the use of an internal complex Su-field (A.15).

$$S_1 \nearrow\overset{F}{}\nwarrow S_2 \Longrightarrow S_1 \longleftarrow\overset{F_1}{} (S_2, S_3) \qquad\qquad (A.15)$$

In order to make the operation of taking off the pack easier it is proposed (prior to submerging into the melt) to introduce a sub-layer, which contains an evaporative

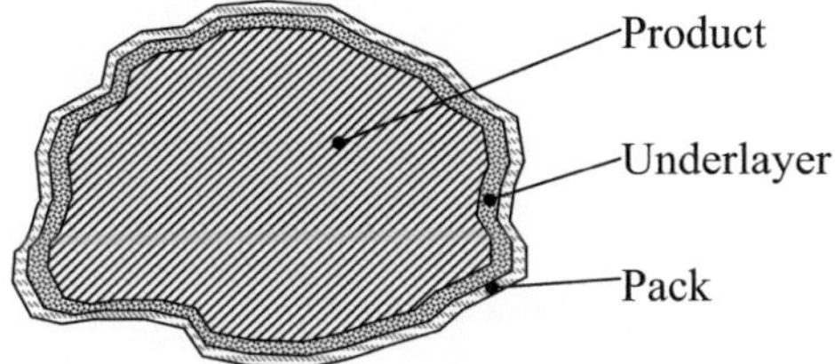

Fig. A.13 Packing a product with a complicated relief surface

substance at low temperature (Fig. A.13). Taking the pack off is performed by heating the pack. Vapor is generated under the pack, which tears the pack.[10]

The underlayer is introduced by submerging the product into a melt of evaporative substance or by applying it using a pulverizer. After that, the product is cooled, for example, by blowing air around it. After that, it is submerged into a polymer melt.

The composition used for the obtainment of the underlayer consists of 50% of an easily volatile solvent of acetone and 50% of polymer composition, used for the obtainment of the main packing film, which prevents the contamination of the bath with melt, intended for application of the main packing film with a heterogeneous composition.

In the model (A.15)

S_1 product
S_2 pack
F force, by which the pack is held on the product
S_3 underlayer containing evaporating substance
F_1 temperature field

Problem 20. Supports of the bridges

Conditions of the problem

In winter time the supports of the bridges are covered with ice and are gradually destroyed. Water gets into the cracks formed in the supports. As a result of freezing ice is expanded and splits pieces off.

It is necessary to eliminate this disadvantage in the simplest and cheapest way.

Analysis of the problem

Condition—prototype

Water S_2 washes the support S_1. Low temperature acts upon water S_2 and water is converted into ice, which spoils the supports.

Su-Field pattern of the problem is described by the formula (A.16).

---

[10] Author's certificate 880,889.

$$
\begin{array}{c}
F_1 \searrow \\
S_1 \longleftarrow\!\!\sim\!\!\longrightarrow S_2
\end{array}
\qquad (A.16)
$$

where

S_1 support of the bridge
S_2 water
F_1 temperature (below zero)

Solution

Problem is described by a Su-Field with a harmful bond. One of the possible solutions is to pass over to the **External Complex Su-field** (A.17).

$$
\begin{array}{ccc}
F_1 \searrow & & F_1 \searrow \\
S_1 \longleftarrow\!\!\sim\!\!\longrightarrow S_2 & \Longrightarrow & S_1 \longleftarrow S_2, S_3
\end{array}
\qquad (A.17)
$$

It is necessary to introduce S_3, which does not enable water to freeze.
Let us use the resources of the system.
Below the surface of the water, especially at the bottom, water is characterized by temperature, which is above zero. The only thing that remains to be done is to transfer this heat to the surface around the support. It is necessary to use a substance with a higher heat conductivity coefficient, for example, copper. Copper tubes are placed around the supports, which are hammered into the ground (bottom). They will transfer heat to the surface and warm water around the supports (Fig. A.14).

S_3 copper tube

In a more detailed way, the model (A.17) could be presented in the following way (A.18).

$$
\begin{array}{ccc}
F_1 \searrow & & F_1 \searrow \\
S_1 \sim S_2 & \Longrightarrow & S_1 \longleftarrow S_2, \quad S_3 \quad S_4 \\
& & \qquad\quad \nwarrow\!\!\nearrow\quad\nwarrow\!\!\nearrow \\
& & \qquad\qquad F_2 \qquad F_3
\end{array}
\qquad (A.18)
$$

where

S_4 warm water (near the bottom of the river)
F_3 temperature above zero, transferred from water S_4
F_2 temperature above zero, transferred from the copper tube S_3

Fig. A.14 Protecting the supports of the bridge

References

1. Altshuller GS (1984) Creativity as an exact science: the theory of the solution of inventive problems. Gordon and Breach, Luxembourg
2. Altshuller GS (1986) To find an idea. Introduction to the theory of inventive problem solving. Nauka, Novosibirsk, 209 c (Russian)
3. Altshuller GS (1988) Little immense worlds: standards for inventive problem solving. A thread in the labyrinth. Karelia, Petrozavodsk, pp 165–231 (Russian)
4. Petrov V (2002) Structural Su-field analysis. Tel-Aviv (Russian). http://www.trizland.ru/trizba.php?id=111
5. Petrov V (2002) Degree of fragmental increase. Tel-Aviv (Russian). http://www.trizland.ru/trizba/pdf-books/zrts-13-droblenie.pdf
6. Petrov V (2002) The pattern of transition to the capillary-porous materials. Tel-Aviv (Russian). http://www.trizland.ru/trizba/pdf-books/zrts-14-kpm.pdf
7. Petrov V (2002) Increase of energy density and information density law. Tel-Aviv (Russian). http://www.trizland.ru/trizba/pdf-books/zrts-16-energo.pdf
8. Petrov V (2018) Structural analysis of systems: Su-field analysis. TRIZ/Vladimir Petrov: Publishing Solutions, 212 p. ISBN 978-5-4493-9970-0 (Russian)

V. Petrov, *Structural System Analysis*, https://doi.org/10.1007/978-3-031-55825-2

Index